4th Polish Mining Congress 2017

AA006719

IOP Conference Series: Earth and Environmental Science
Volume 174

Krakow, Poland
20 - 22 November 2017

Editors:

Piotr Banka
Marcin Lutynski
Marta Matyjaszek

ISBN: 978-1-5108-6747-5
ISSN: 1755-1307

Printed from e-media with permission by:

Curran Associates, Inc.
57 Morehouse Lane
Red Hook, NY 12571

CURRAN ASSOCIATES INC. proceedings.com

Some format issues inherent in the e-media version may also appear in this print version.

Printed by Curran Associates, Inc. (2018)

Institute of Physics
Dirac House, Temple Back
Bristol BS1 6BE UK

Phone: 44 1 17 929 7481
Fax: 44 1 17 920 0979

techtracking@iop.org

Additional copies of this publication are available from:

Curran Associates, Inc.
57 Morehouse Lane
Red Hook, NY 12571 USA
Phone: 845-758-0400
Fax: 845-758-2633
Email: curran@proceedings.com
Web: www.proceedings.com

Table of contents

Volume 174

4th Polish Mining Congress-Session: Human and environment facing the challenges of mining

20–22 November 2017, Krakow, Poland

Accepted papers received: 25 June 2018
Published online: 5 July 2018

Preface

Papers

IOP Conf. Series: Earth and Environmental Science **174** (2018) 011001 doi:10.1088/1755-1315/174/1/011001

Preface

Dear Ladies and Gentlemen,

Between 20th and 22th of November 2017 in Cracow (Poland) took place the 4th Polish Mining Congress. Congress was held under the slogan:

„Mineral resources in Poland's economic strategy – how to mine, how to acquire, how to save in the era of sustainable development?"

Over 600 participants attended the congress - mines employees, companies working for the mining industry, scientists, universities' representatives, scientific research centres and units.

The following subjects were within the scope of this congress:

- mining industry and sustainable development policy
- security of energy and raw materials in Poland
- IT and modern solutions in mines – implementation and usage
- mineral resources management

The conference series contains only selected articles. Most of them were presented during the session „Human and environment facing the challenges of mining".

All articles were peer reviewed through process administrated by the Editors.

Piotr Dańka   Marcin Lutyński   Marta Matyjaszek

IOP Conf. Series: Earth and Environmental Science **174** (2018) 011002 doi:10.1088/1755-1315/174/1/011002

Peer review statement

All papers published in this volume of *IOP Conference Series: Earth and Environmental Science* have been peer reviewed through processes administered by the proceedings Editors. Reviews were conducted by expert referees to the professional and scientific standards expected of a proceedings journal published by IOP Publishing.

IOP Conf. Series: Earth and Environmental Science **174** (2018) 012001 doi:10.1088/1755-1315/174/1/012001

The induced seismicity predicting based on changes in the specific energy of elastic strain of tremor-prone rock layers

Piotr Bańka

Faculty of Mining and Geology, Silesian University of Technology, Akademicka 2, 44-100 Gliwice, Poland

piotr.banka@polsl.pl

Abstract. The paper presents the method of predicting changes of seismic hazards in the runways of the designed excavation workings. The presented method uses correlations between the level of observed induced seismicity and the estimated energy states of the rockmass. It is possible to make predictions of time-space changes in the number and energy expenditure of the tremors accompanying the planned mining workings and the changes in induced seismicity as a function of time. The presented results of the calculations performed for the selected area of the hard coal mine revealed the existence of a correlation between the energy changes, taking place in the rockmass and the observed level of seismicity. These dependencies can be used to predict seismic hazards of the planned works in the same part of deposits.

1. Introduction

The seismic hazard belongs to the most dangerous natural hazards in Polish hard coal mines. In the next few years, this threat will not decrease, despite the possible further reduction of hard coal mining. The persistence of the observed level of danger (or even its increasing) will be caused by the necessity of reaching for residual resources lying at ever greater depths. Mining work carried out in such conditions usually generate high-energy tremors, they are also accompanied by a high rockburst threat. An appropriately early prediction of the seismic hazard on the runway of the designed works allows for the verification of the project in terms of minimizing the risk's level, as well as for appropriate planning of prevention measures. In the absence of an appropriate network of excavations or the retention of tremor-prone rock layers at a large distance from existing workings, it becomes necessary to use analytical methods to assess the state of danger. These methods, despite the limitations resulting from the necessity to adopt many simplifying assumptions and due to the high variability of strength and deformation parameters of rock layers, enable preliminary, approximate predictions of the potential seismic hazard and rockbursts at the stage of mining works design.

The searching for seismic relationships with geological and mining conditions is mostly empirical. Among the large number of research in this area, it is possible to mention, for example, works from recent years [1 ÷ 4]. The level of recorded seismic activity induced by conducted mining works can be determined by the deformation processes occurring in the rockmass. The possibility of taking into account the indicators describing the deformation processes of rockmass layers and the terrain surface to characterize the induced seismicity level was pointed out in numerous works, including [5 ÷ 10]. In many works, among others [11 ÷ 13], was showed a correlation relationship between the estimated energy changes in the rockmass and the size of the induced seismicity.

During the many years of research conducted at the Institute of Mining of the Silesian University of Technology, was discovered a correlation relationship between the estimated energy changes in the rockmass and the level of induced seismicity. The description of energy changes concerns the elastic strain energy, accumulated in the rockmass as a result of disturbing it, it says nothing about its dissipation and transformation, and consequently does not specify the amount of energy released in the process of damaging a given volume of rocks. In this process, the potential elastic strain energy passes into other types of energy, including the kinetic energy of elastic waves associated with the seismic energy of recorded tremors [14]. In this way, potential elastic strain energy changes translate into the level of induced seismicity. This justifies the searching for relationships between the estimated changes in potential energy and the seismic energy of tremors.

The article discusses the method of using calculated changes of specific energy to determine the magnitude of the number and energy expenditure of tremors; an example of the prediction of the level of seismicity induced in a strongly threatened with tremors area of a hard coal mine is presented.

2. Relationship between calculated changes of the specific energy and the observed seismicity level

The specific energy of elastic strain Φ [J/m^3] is determined by a known Clapeyron equation:

$$\Phi = \frac{1}{2} T_\sigma T_\varepsilon, \tag{1}$$

where:

T_σ – tensor of stress state, MPa;

T_ε – tensor of strain state.

The analytical methods may be used to estimate energy changes taking place in tremor-prone rock layers, e.g. a method based on the solution of the dislocation boundary value problem of the spatial theory of elasticity given by H. Gil [15], determining the distribution of stresses and strains in the half-space around the rectangular void. For this purpose it is also possible to use a much more modern tool, which is the finite difference method (e.g. the FLAC3D program).

Analytical methods allow to calculate energetic changes without differentiating mechanical parameters and geometry of considered rock layers. These methods are very numerically effective, allow calculation of energy changes in any time intervals and areas, so that instantaneous or local extremes are not omitted – large energy changes associated with configuration and time of operation. This is very important because the values of potential energy undergo strong changes in the period when the rockmass is covered by the influence of the progressing exploitation front. However, these methods require a strong idealization of the rock medium (continuous, homogeneous, isotropic, linear elastic center). Therefore, a typical comparative approach is required (prediction precedence by computational tests for the appropriate range of the already performed operation) to determine the values of parameters characterizing the rockmass, at which the forecast results will be the closest to reality.

Work [16] analyzed the possibility of increasing the accuracy of the description of energy changes occurring in the rockmass by using the finite difference method, which does not require such a strong idealization of the center. Numerical methods allow using practically any complex description of geomechanical processes occurring in the rockmass. However, there are two major limitations in taking full advantage of these opportunities. The first of them, which is gradually losing its importance, is the limited computing power of computers. The second obstacle is the ignorance of the exact geological structure of the rockmass and the geomechanical parameters of the rocks, often causing that the large possibilities of numerical methods are not fully used. As shown by the results of the test calculations [16], with the appropriate selection of parameters of the analytical method, it is possible to obtain comparable results of calculations of energy changes to those obtained by the numerical method.

The nature of changes in the value of specific energy and induced seismicity are similar. By way of example, figure 1 shows an example of the formation of seismicity induced by mining works carried out in the area of the impact of the stopped operation edge (a), as well as the remains of the deposit left in the adjacent seam (b). There were used data on induced seismicity published in the paper [17]. Value of seismic activity E/W was calculated by dividing the sum of registered during one day energy of tremors and coal output (in the same period). The graphs also show changes in specific energy values calculated using the analytical method and the FLAC3D program.

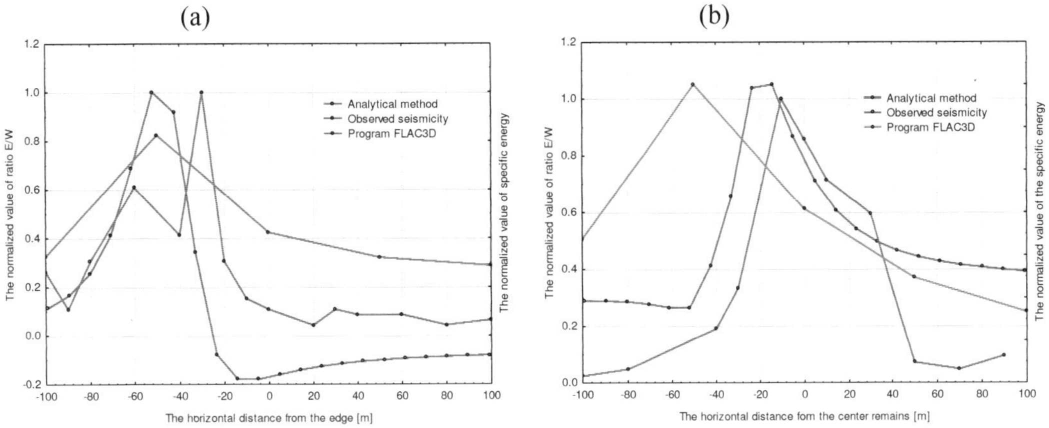

Figure 1. Changes of the values of seismicity and specific energy during the working of longwall under the mining edge (a) and the residue (b) in the overlying seam.

Tremors induced by mining exploitation occur both on the runways and in the goafs of the operating longwalls (during loading and unloading of the deformed rock layer). Therefore, it is reasonable to compare the magnitude of seismicity induced both with calculated increases and decreases in specific energy. The dissipation of accumulated energy is different when the rocks are not damage, and different when the rocks have already been damaged (as a result of the impact of previously conducted mining works). Considering the observations above, the adopted regression models include the calculated increases (decreases) of specific energy with the separation of the taking place before the hypothetical damage of the rock (W effort rate <1) and after it (W effort rate ≥ 1). In order to determine whether rocks have been damaged, the Hock-Brown generalized condition is used [18]. Energy changes taking place in one or several rock layers are taken into account.

The method of regression analysis was used for searching the relationships between energy states of the rockmass and the level of induced seismicity. This method allows to approximate the functional dependence with another, simpler function, such as a polynomial that contains the appropriate variables and approximates the function sought in certain limited intervals [19]. Space-time changes in seismicity and the formation of these changes over time were analysed.

2.1. Model of seismic changes in a space-time profile

In this model, the number of tremors per surface unit or the energy expenditure of tremors per unit of surface within the adopted time interval determines the expression:

$$E_{cal\,l} = a_0 + \sum_{k=1}^{Nl} \sum_{j=1}^{4} a_{kj} f_{jkl} + \varepsilon \qquad for \quad l = 1 \ldots Np \tag{2}$$

where:

$E_{cal\,l}$ - considered seismic rate calculated in point l,

Nl - number of tremor-prone rock layers,

Np - number of calculation points,

a_0, a_{kj} - parameters of the regression model,
Nt - number of unit time intervals,
ε - random component,
$f_{1kl} = \sum_{i=1}^{Nt} \Delta\Phi^{+}_{i(W<1)}$ – the sum of instantaneous increases of specific energy calculated in point l for the k-th rock layer, when the effort rate W <1,
$f_{2kl} = \sum_{i=1}^{Nt} \Delta\Phi^{-}_{i(W<1)}$ – the sum of instantaneous decreases in specific energy calculated in point l for k-th rock layer, when the effort rate W <1,
$f_{3kl} = \sum_{i=1}^{Nt} \Delta\Phi^{+}_{i(W\geq1)}$ – the sum of instantaneous increases of specific energy calculated in point l for the k-th rock layer, when the effort rate W≥1,
$f_{4kl} = \sum_{i=1}^{Nt} \Delta\Phi^{-}_{i(W\geq1)}$ – the sum of instantaneous decreases in specific energy calculated in point l for the k-th rock layer, when the effort rate W≥1.

2.2. Model of changes in seismicity over time
In this model, the number of tremors per unit of time or the sum of seismic energy of tremors in a unit of time is determined by the expression:

$$E_{cal\,i} = a_0 + \sum_{k=1}^{Nl} \sum_{j=1}^{4} a_{kj} g_{jkl} + \varepsilon \quad for \quad l = 1 \ldots Nt \tag{3}$$

where:
$E_{cal\,i}$ - considered seismic rate calculated in the i-th time interval,
$g_{1ki} = \sum_{l=1}^{Np} \Delta\Phi^{+}_{l(W<1)}$ – calculated in the i-th time interval, for the k-th rock layer of the sum of instantaneous increases of specific energy, when the effort rate W <1,
$g_{2ki} = \sum_{l=1}^{Np} \Delta\Phi^{-}_{l(W<1)}$ – calculated in the i-th time interval, for the k-th rock layer, the sum of instantaneous decreases in specific energy, when the effort rate W <1,
$g_{3ki} = \sum_{l=1}^{Np} \Delta\Phi^{+}_{l(W\geq1)}$ – calculated in the i-th time interval, for the k-th rock layer of the sum of instantaneous increases of specific energy, when the effort rate W≥1,
$g_{4ki} = \sum_{l=1}^{Np} \Delta\Phi^{-}_{l(W\geq1)}$ – calculated in the i-th time interval, for the k-th rock layer of the sum of instantaneous decreases in specific energy, when the effort rate W≥1.
Other markings – as above.

In addition, the following limitations have been accepted:

$$a_{kj} \geq 0 \text{ for } k = 1 \ldots Nl, j = 1 \ldots 4 \tag{4}$$

To determine the parameters of the model (minimization of the objective function - L1 or L2 norm of the vector of differences calculated and observed values of the number or the energy expenditure of tremors) the evolutionary difference algorithm [20] is used.

In the conducted research the following indices characterizing the level of seismicity of exploitation regions were adopted. During the analysis of space-time distributions:

- number of tremors per surface unit [$1/(10^4\ m^2)$],
- energy expenditure of tremors per unit of volume [J/m^{3-}].

In turn, when analysing the distributions of recorded seismicity over time, the following indices was calculated:

- number of tremors registered in a unit time interval, e.g. 7 days [1/week],
- total energy of tremors that occurred in the analysed time period, e.g. 7 days [J/week].

3. Correlations between number and energy expenditure of tremors and energy changes

The analysis of the impact of energy changes occurring in the rockmass on the level of induced seismicity was carried out for the area of works that took place in conditions of high risk of tremors and rockbursts in one of the hard coal mines located in Upper Silesia Coal Basin, Poland. The exploitation was carried out in the areas of two longwalls 002 and 003 in the upper layer of seam 504, the thickness of which varies from 2.8 m to 7.0 m (the height of the longwall was up to 3.4 m). The depth of the deposit is about 850 m.

In the area of the presented longwall excavations, there are exploitation edges in the above-declining seams, including 413 at a distance of about 270 m, 416 at a distance of about 160 m, 418 at a distance of about 130 m and 502 at a distance of about 70 m from seam 504.

Mining work in longwalls 002 and 003 induced 5000 tremors with a total energy of about $4x10^7$ J. The strongest phenomenon occurred during operating the longwall 002 and reached the energy of $6x10^6$ J. There were registered 3816 tremors with energy of 10^2 J, 964 with energy of 10^3 J, 194 of energy of the order of 10^4 J. There occurred 26 high energy phenomena, including 20 with the energy of 10^5 J and 6 with energy of the order of 10^6 J.

The period of conducting the works was divided into two sub-periods:

- 04.2009 ÷ 12.2012 - using data on registered seismicity in this time interval and calculated values of indicators characterizing energy changes, estimation of regression model parameters was carried out,
- 05.2014 ÷ 05.2015 - for which, using the determined regression equations, calculations of changes in the number and density of energy of tremors were performed. The predicted changes in seismicity were compared with those observed in the course of works in the longwall 003.

3.1. Description of changes in seismicity in the space-time profile

3.1.1. Analysis of energy density of tremors. The results of calculations of the regression model parameters (2) based on the data observed during the operating longwall 002 are given in table 1. The applied model incudes a constant a_0. It is worth to note, that very similar results were obtained for a model that does not contain this constant [21].

Table 1. Results of computation of regression model (2) parameters – taking into account tremors registered during the working of longwall 002, tremors energy density analysed [J/m^3].

Layer number	Depth [m]	Values of the regression model parameters; a_0 = -0.71			
		a_{1j}	a_{2j}	a_{3j}	a_{4j}
1	400	122.386	0	0	32.193
2	500	97.636	11.5	4.923	18.195
3	750	10.984	31.148	0.042	9.279
4	830	8.814	0	0	1.228

The determined regression equation enabled relatively accurate reproduction of changes in the energy density value of tremors observed on longwall runway 002. Figure 2a shows the observed, whereas in figure 2b, there is presented the energy density distribution calculated from the determined equation prepared for the longwall operating period 002. The increased values of the energy expenditure of tremors within the range of the impact of the edges produced in the seam 416, and also in the seam 502 are visible. Unlike the regionalization, the extreme values of the observed energy expenditure of tremors could not be recovered. The value of the mean square error is 1.93 J/m^3, the value of the linear correlation coefficient is 0.8 - the coefficient of determination is 0.64, i.e. 64% of the observed variation of the energy density of tremors is explained by the determined equation.

(a) (b)

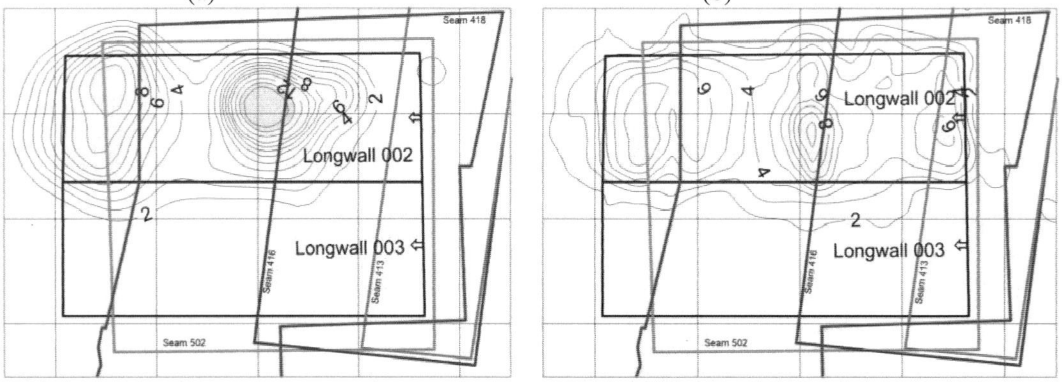

Figure 2. Distributions of observed (a) and calculated (b) energy density of tremors [J/m^3] during the working of longwall 002.

The regression equation (2) has been used to predict changes in energy density of tremors on longwall 003. The result of the energy expenditure prediction for tremors made for the second stage of works, shown in figure 3b, was compared with the observed distribution of tremors energy density, which is shown in figure 3a.

(a) (b)

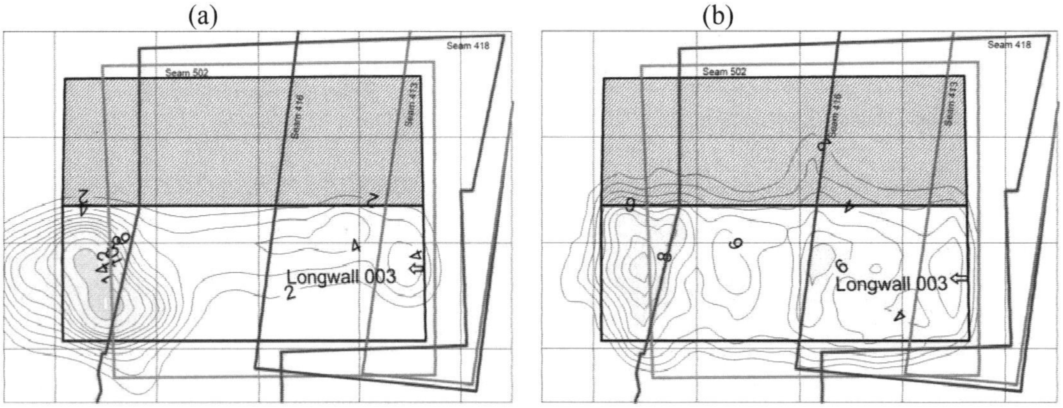

Figure 3. Distributions of observed (a) and predicted (b) energy density of tremors [J/m^3] during the working of longwall 003.

The estimated model allowed a relatively accurate prediction of the induced seismicity on longwall 003. The value of the mean square error is 2.1 J/m^3, the value of the linear correlation coefficient between the observed and predicted values is 0.7 (value of the determination coefficient of about 0.5). In particular, the most endangered area, which was the final section of the longwall, was aptly identified. It is important that the results indicated the possibility of a higher value of the maximum energy expenditure of tremors than that observed during the guiding of the longwall 002. However, it was not large enough to suggest on this basis the earlier stop of the longwall run. In general, the prediction for the majority of area of longwall 003 was a bit over-heated. This fact can be explained, except the inevitable mistakes resulting from the simplicity of the model used, also with a very wide range of rockburts's prevention.

3.1.2. Analysis of the number of tremors. The results of the calculation of the regression model parameters based on the data observed in the first period - during the operating of the longwall 002 are given in table 2.

Table 2. Results of computation of regression model (2) parameters – taking into account tremors registered during the working of longwall 002, analysing the number of tremors per 10^4 m^2.

Layer number	Depth [m]	Values of the regression model parameters; a_0 = -8.006			
		a_{1j}	a_{2j}	a_{3j}	a_{4j}
1	400	382.395	0.000	0.000	209.807
2	500	298.478	458.351	16.391	316.949
3	750	130.647	399.711	0.000	261.143
4	830	368.690	0.000	0.000	0.000

Analogically, as in the previous case, the determined regression equation made it possible to reproduce the changes in the number of tremors that occurred on longwall 002 in a correct way. Figure 4a presents the observed, while in figure 4b, the distribution of number of tremors calculated for the period of 002 operating longwall. The value of the mean square error is 24.1x10^{-4} m^{-2}, the value of the linear correlation coefficient is 0.8 - the determination coefficient is 0.58, i.e. 58% of the observed variation in the number of tremors is explained by the determined equation.

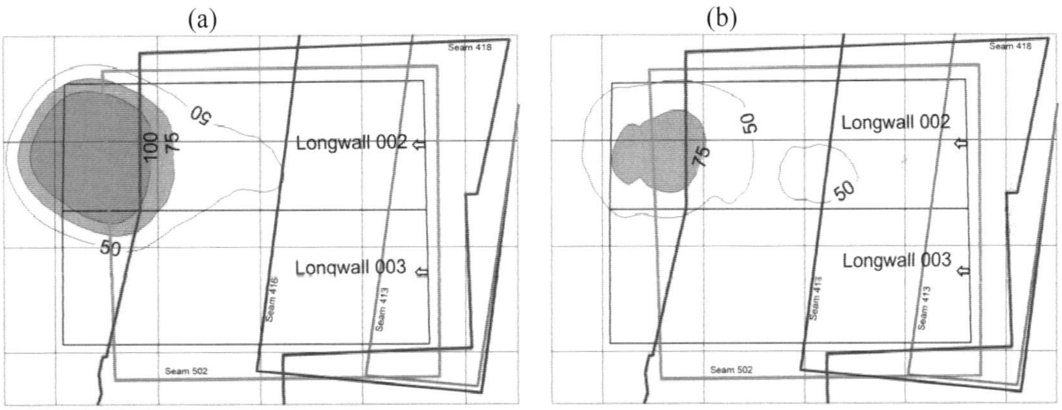

Figure 4. Distributions of observed (a) and calculated (b) number of tremors [x10^{-4} m^{-2}] during the working of longwall 002.

The determined regression equation was used to make a prediction of changes in the number of tremors on longwall runway 003. The result of the seismic prediction prepared for the second stage of works, shown in figure 5b, was compared with the observed distribution of number of tremors, which is shown in figure 5a.

Similarly to the analysis of the energy expenditure of tremors, the determined equation was able to accurately indicate the most endangered region, which was the final section of the longwall 003. The value of the mean square error is 46.1x10^{-4} m^{-2}, the value of the linear correlation coefficient between observed and predicted values is 0.6 (value of the determination coefficient of about 0.4).

3.2. Description of changes in seismicity over time

In some cases, it may be useful to use the (3) model, which allows description of induced seismicity changes in time. The use of this model is particularly appropriate in cases where data of registered tremors come from areas that are difficult to observe and therefore the location of tremors epicenters can be loaded with significant errors. During the estimation of parameters of equation (3) errors of the location of occurring tremors do not affect the value of regression coefficients. Similarly, as during the analysis of space-time changes in seismicity, the results of calculations for two indicators characterizing the level of observed seismic hazard were given: logarithm of the sum of tremors' energy that occurred in 7 days (table 3) and the number of tremors that occurred in the same, 7-date period (table 4).

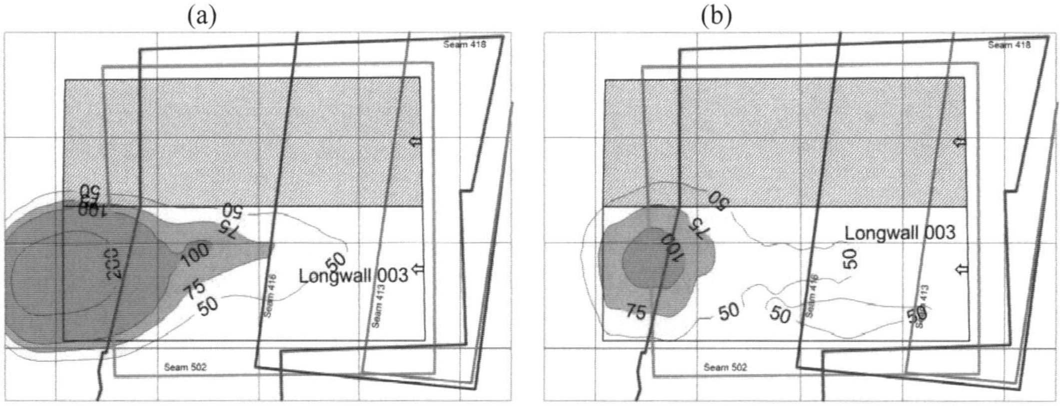

Figure 5. Distributions of observed (a) and predicted (b) number of tremors [x10^{-4} m^{-2}] during the working of longwall 003.

Table 3. Results of computation of regression model (3) parameters – taking into account tremors registered during the working of longwall 002, analysed the total energy of tremors which occurred in 7-days periods of time [J/7 days].

Layer number	Depth [m]	Values of the regression model parameters;			$a_0 = 4.560616$
		a_{1j}	a_{2j}	a_{3j}	a_{4j}
1	400	0	0	0	5.172426
2	500	0	0	0	0
3	750	0	0	0	0.139916
4	830	0	0	0	0

Table 4. Results of computation of regression model (3) parameters – taking into account tremors registered during the working of longwall 002, analysed the number of tremors which occurred in 7-days periods of time [1/7 days].

Layer number	Depth [m]	Values of the regression model parameters;			$a_0 = 0$
		a_{1j}	a_{2j}	a_{3j}	a_{4j}
1	400	0	1.57E-05	0	253.4174
2	500	0	0	0	0
3	750	0	126.5644	0	0
4	830	28.17377	0	0	0

It is noteworthy that the energy processes occurring in the second tremor-prone layer, lying at a depth of 500 m, do not affect the observed changes in time-induced seismicity. The results of the reproduction of the observed number of tremors and their total energy in the weekly time intervals, for the longwall operating period 002 (dashed line) and the predictions of these indicators for the longwall operating period 003 (continuous line) are shown in figure 6. In addition, in these drawings a distribution of changes in observed seismicity over time is reported (thick, continuous line).

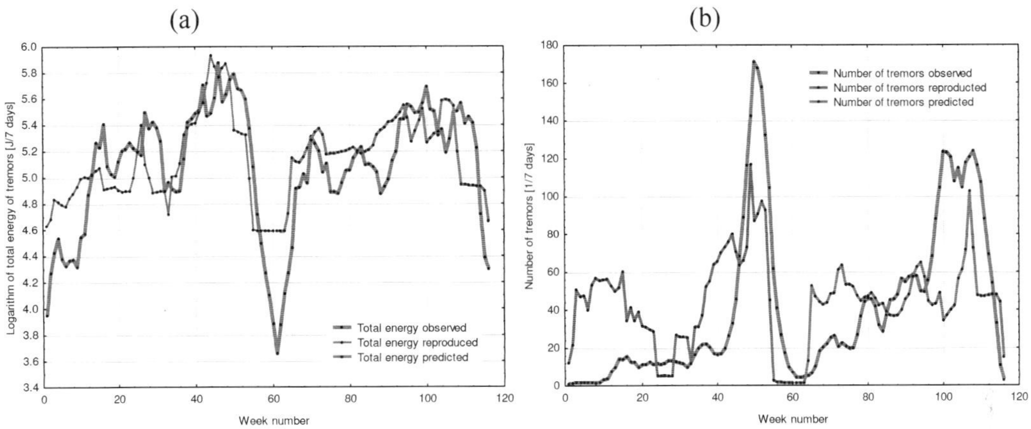

Figure 6. Distributions of observed, computed and predicted logarithms of total energy of tremors [J/7 days] (a) and number of tremors [1/7 days] (b) registered in 7-days periods of time.

As a result of the calculations made for the seismicity described by the logarithms of the sum of tremors energy in unit time intervals (figure 6a), the following estimates of regression accuracy were obtained: for longwall guide period 002 (estimation of model parameters) the mean square error value is 0.33 J/7 days, in turn the value of the linear correlation coefficient between observed and predicted values is 0.8 (value of the coefficient of determination about 0.6); for the second period, operating longwall 003, i.e. for the time interval in which the prediction of changes in seismicity was made, the mean square error value is 0.34 J/7 days, while the value of the linear correlation coefficient between observed and predicted values is 0.7 (value of the determination coefficient about 0.5).

If the level of induced seismicity was characterized by specifying the number of phenomena that occurred in a unit time interval (figure 6b), the estimation of the regression accuracy was as follows: for longwall operating period 002 the mean square error value is 37 tremors/7 days and the value of the linear correlation coefficient is 0.6 (value of the coefficient of determination about 0.4); for the period of prediction of changes in seismicity, the mean square error value is 35 tremors/7 days, while the value of the linear correlation coefficient is 0.5 (the value of the determination coefficient is about 0.3). Despite the low values of the coefficient of determination and the big values of the mean square error, it is worth noting that the period of the greatest seismic threat accompanying the works run in the longwall 003 and its level was accurately predicted.

4. Summary

The calculations carried out for many regions allowed to determine the existence of correlative relationships between parameters characterizing the course of energy processes occurring in tremor-prone rock layers and the level of induced seismicity. These dependencies can be used to assess the level of seismicity on runs of designed works in the same area, with accuracy sufficient for mining practice.

The results of previous studies indicate that the most accurate results of prognostic calculations should be expected during works at a small distance from tremor-prone layers and within the range of the impact of near exploitated edges. In the case of hard rock layers or past exploitation at large distances from the selected seam, the inference regarding the size of seismicity on the planned longwall runs is subject to a greater error. The presented method may be a supplement to the methods used to assess the status of a seismic hazard. This method can be used primarily in those areas where the seismicity level is determined by the configuration of edges and remains in the exploited seam as well as in neighboring seams, and to a lesser extent, the variable geological structure of the rockmass. The method is local, it means that prediction with the use of the determined regression equation should be performed only for the region from which the data used to estimate the model parameters came from.

References

[1] Abdul-Wahed M K, Al Heib M and Senfaute G 2006 Mining-induced seismicity: seismic measurement using multiplet approach and numerical modeling *International Journal of Coal Geology* **66/1** pp 137–47

[2] Babenko E V 2010 Prediction of mining induced seismicity around moving longwall *Transactions of UkrNDMI NAN Ukraine* **7** pp 76–85

[3] Kudela J, Kurnik S and Rusinek J 2009 Ocena wpływu koncentracji wydobycia na poziom zagrożenia sejsmicznego w partii VII pokładu 209, KW SA KWK „Piast" *Prace Naukowe GIG* **4/2** pp 169–79

[4] Majcherczyk T and Olechowski S 2011 Parametry frontu eksploatacyjnego a zagrożenie sejsmiczne w rejonie W1 Kopalni Węgla Kamiennego „Rydułtowy-Anna" *Górnictwo i Środowisko* **4/2** pp 232–42

[5] Białek J, Drzęźla B and Jaworski A 1992 Próba ustalenia zależności funkcyjnych pomiędzy przebiegiem deformacji górotworu w czasie a energią sejsmiczną dla warunków KWK Rydułtowy *Publ. Inst. Geophys. Pol. Acad. Sc.* **M-16** pp 279–87

[6] Bańka P 2004 Determination of potential level of rockburst hazard based on the result of analytical predictions of stress distributions and the level of induced seismicity *Acta Geodynamica et Geomaterialia* IRSM AS CR **1** pp 19–26

[7] Głowacka E 1991 *Ocena zagrożenia sejsmicznego górotworu z uwzględnieniem przebiegu eksploatacji* Thesis Instytut Geofizyki PAN Warszawa

[8] Goszcz A 1988 Wpływ gradientu prędkości obniżania się powierzchni pod wpływem robót górniczych na stan zagrożenia wstrząsami górniczymi *Zeszyty Naukowe AGH seria Górnictwo* **141** pp 63–9

[9] Jarmużek J, Siewierski S and Wilczyński W 1987 Wpływ powierzchni wybranego złoża oraz deformacji terenu na aktywność sejsmiczną górotworu na przykładzie ZG Rudna *Zeszyty Naukowe AGH seria Górnictwo* **129** pp 67–73

[10] Wanior J 1992 *Metoda prognozowania wstrząsów i tąpnięć w oparciu o wyniki pomiarów geodezyjnych* (Częstochowa: PTPNOZ)

[11] Bańka P and Jaworski A 2007 Analitycznie symulowane zmiany stanów naprężeniowo-energetycznych górotworu w kontekście potencjalnego zagrożenia tąpaniami robót górniczych *Mechanizacja i automatyzacja w górnictwie* **9** pp 17–27

[12] Białek J, Bańka P and Jaworski A 1999 Analityczne prognozowanie zmian energii sprężystej warstw skalnych *Zeszyty Naukowe Politechniki Śląskiej s. Górnictwo* **239** pp 93–102

[13] Jaworski A 2004 Description of induced seismicity level basing on analytically calculated changes of elastic energy in rock layers subjected to deformation *Acta Geodynamica Et Geomaterialia* **1** pp 7–17

[14] Salamon M D G 1984 Energy Considerations in Rock Mechanics: Fundamental Results *IS.Afr.Inst.Min.Metal.* **84/8** pp 951–1099

[15] Gil H 1991 *The Theory of Strata Mechanics* (Warszawa: PWN)

[16] Bańka P 2013 *Modelowanie zmian sejsmiczności indukowanej na podstawie szacowanych stanów energetycznych górotworu* (Gliwice: Wydawnictwo Politechniki Śląskiej)

[17] Gerlach Z 1990 Empiryczne modele przewidywania stanu zwiększonego zagrożenia tąpaniami

w oparciu o wyniki sejsmologii górniczej Thesis AGH Kraków
[18] Hoek E, Carranza-Torres C and Corkum B 2002 Hoek-Brown Failure Criterion – 2002 edition *Proc. NARMS-TAC Conference* Toronto pp 267–73.
[19] Draper N R and Smith H 1966 *Applied Regression Analysis* (New York: Wiley)
[20] Storn R and Price K 1997 Differential Evolution – A simple and efficient heuristic for global optimization over continuous spaces *Journal of Global Optimization* **11** pp 341–59
[21] Bańka P, Chmiela A, Menendez Fernandez M, Fernandez Muniz Z and Sanchez A B 2016 Predicting changes in induced seismicity on the basis of estimated rock mass energy states *Int. J. Rock. Mech. Min. Sci.* **95** pp 79–86

Forecasting of stresses in overhead power lines running through area affected by the mining damage

Sergiusz Boron

Faculty of Mining and Geology, Silesian University of Technology, Akademicka 2, 44-100 Gliwice, Poland

sergiusz.boron@polsl.pl

Abstract. The article presents the methodology of determining the impact of subsidence resulting from underground mining activity on the operation of overhead power lines. Exemplary results of calculations of stress in conductors has been presented. Also the method of defining the influence of factors that can mitigate the effects of land deformation, such as inclination of insulator chains and elastic deformation of overhead line towers and their crossarms, has been discussed.

1. Introduction

The planning of underground mining should take into account the impact of subsidence on buildings and technical infrastructure, including power lines. Forecasting of this impact requires knowledge of surface deformation indicators in the analyzed region in particular stages (years) of planned exploitation. Deformations may cause hazards to power lines that run through the area affected by mining damage. These hazards include the possibility of breaking the line due to exceeding the tensile strength of conductors, damaging the transmission support structures and excessive proximity of the power cables to the ground or other crossed objects. The paper presents the method of forecasting of the subsidence impact on power lines, assessment of the influence of factors that can mitigate the effects of deformation on conductors' stresses and conclusions resulting from exemplary calculations.

2. Theoretical foundations and assumptions used in calculations

The basic guidelines for the design and construction of overhead power lines can be found in the standard [1], which was withdrawn in 2003 without substitution, however most of power lines in Poland have been constructed in accordance with the requirements specified there. This standard comprises, among others, general recommendations regarding the operation of the lines running through the mining damage areas, however, the methodology for determining (calculating) the impact of subsidence on the existing line has not been defined in this standard.

During assembly of the line, the conductor should be so stretched that, under operating conditions, no maximum permissible stress is exceeded at any point of the line. The stress should not exceed elastic limit in two conditions causing the occurrence of maximum stress:

- temperature -25 °C without the ice loading,
- temperature -5 °C with the ice loading.

In the case of short spans, the greatest stress is caused by frost (-25 °C) whereas for longer spans, higher stresses are caused by ice loading (-5 °C). In exceptional weather conditions that occur rarely and last for a short time, the conductor may be covered by double ice loading (with a weight twice as large as normal rime). The stress under these conditions may be greater than the allowable normal stress, but it shouldn't exceed permissible extreme stress. The second quantity subject to the normative requirements is the ground clearance of the conductors. The distance between conductors and the ground (or crossed objects) depends on the sag defined as the vertical distance between the lowest point of the conductor and the straight line connecting the suspension points of the line. The maximum sag may occur for the conductor with ice loading at -5 °C or at the maximum operating temperature.

The subsidence above the underground operation may include horizontal and vertical displacements as well as ground tilt changes at supports location. Therefore, the increase or decrease of the span length due to land deformation should be taken into account. The standard [1] comprises guidelines and recommendations for designing overhead lines in mining damage areas, but regardless of this, in the case of large deformations, exceeding the permissible stresses in the conductors is possible. Also increased sag may cause excessive proximity of conductors to the ground or other objects [2].

The basic assumption when analyzing the mechanics of conductor is to consider the change in the span length as an equivalent to the length change of the conductor with the opposite sign [2, 3]. Thus, increasing the span length by Δa a is regarded as shortening the length of the conductor by Δa without change of the span length and vice versa.

The factors characterizing the terrain deformation include (figure 1):

- tilt of the terrain T (‰),
- horizontal deformation ε (‰),
- radius of deformation curvature r (m) or its reciprocal – curvature (m^{-1}),
- vertical displacement (lowering of the terrain) W (m),
- horizontal displacement u (m).

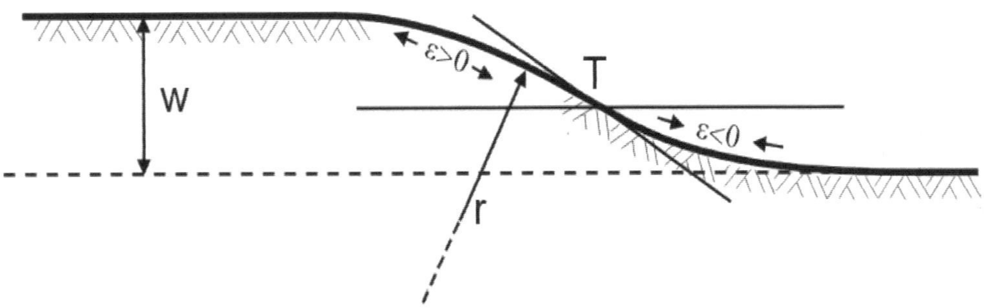

Figure 1. An example of terrain deformation and its parameters.

Only horizontal displacements and tilt at the support location can be considered as significant factors causing changes of span length in the conductor stress calculation. In most cases the change of the span length caused by the lowering the support can be omitted due to relatively small values of vertical displacements in comparison to the span length. In addition, the lowering of the area on which the support stands, usually also causes the lowering of objects under the overhead line, therefore it shouldn't cause the significant increase of the conductors' proximity to these objects.

The conductor suspended between supports is arranged in the shape of catenary curve described by the hyperbolic function. In the practical calculations, when expanding this function into a power series, only first two terms can be taken into account [4], so the hyperbolic function can be replaced by parabolic one. The stress in the conductor can be calculated from the equation of state of a conductor, in the form [2]:

$$\sigma^3 - A'\sigma^2 - B = 0 \tag{1}$$

where:

$$A' = \sigma_1 - \frac{a^2 g_1^2}{24\beta \cdot \sigma_1^2} + \frac{\alpha}{\beta}(t_1 - t) + \frac{\Delta a}{\alpha\beta} \tag{2}$$

$$B = \frac{a^2 \cdot g_0^2}{24\beta} \tag{3}$$

$$\sigma_1 = \sigma_0 - \frac{a^2 g_1^2}{8\sigma_0} \tag{4}$$

The following symbols have been adopted in the above equations:
σ - horizontal stress of a conductor (MPa),
σ_0 – stress in initial (designing) conditions (MPa),
σ_1 - horizontal stress in initial conditions (MPa),
g_1 - volumetric load in initial conditions (MN/m^3),
g_0 - volumetric load in given conditions (load with the mass of the conductor g_n, load with the mass of conductor with ice loading g_s or load with the mass of conductor with double ice loading g_{sk}) (MN/m^3),
a – initial span length (m),
Δa – change in the span length due to horizontal displacement and supports inclination (m),
α - conductor's thermal expansion coefficient (K^{-1}),
β - conductor's elastic elongation coefficient (m^2/N),
t_1 – temperature in initial conditions, (°C),
t – temperature at which the stress σ is calculated, (°C).

Knowing the stress in the conductor under given conditions, the corresponding sag is calculated from the expression:

$$f = \frac{a^2 \cdot g_0}{8\sigma} \tag{5}$$

3. Sample calculations of conductor's stress and sag

As an example of the application of the above methodology, the results of calculations of the selected span of the actual 110 kV overhead line crossing the area with planned underground exploitation has been presented (table 1). The design span length a=218.4 m, the calculations refer to the AFL-6-240 conductor suspended at a height of 23.6 m. Both supports forming the span have suspension (hanging) insulators. Other data necessary for calculations were adopted in accordance with the design and assembly documentation. The planned exploitation covers a period of 16 years and the second column of table 1 presents the changes in the span length caused by terrain deformation in subsequent years (parameters determining deformation are calculated on the basis of operational plans, depth, thickness and slope of extracted beds, method of roof control etc.). Depending on the span length, the conditions under which the stress takes the maximum value have been determined (A – ice loading at -5°C, B – temperature -25°C without ice loading). Following columns of table 1 include the initial (design) conductor stress σ and corresponding sag f (under conditions A or B), stress σ_{sk} and sag f_{sk} for double ice loading as well as stress σ_{40} and sag f_{40} at 40°C.

Table 1. Results of calculations of maximum stress and sag in the exemplary span of the 110 kV overhead line

Year	Δa	Conditions	σ	f	σ_{sk}	f_{sk}	σ_{40}	f_{40}
–	(m)	–	(MPa)	(m)	(MPa)	(m)	(MPa)	(m)
1	0.13	A	71.87	5.48	96.97	5.99	34.22	6.01
2	0.22	A	79.70	4.94	105.33	5.52	37.86	5.43
3	0.17	A	75.21	5.23	100.60	5.78	35.76	5.75
4	0.24	A	82.14	4.79	107.86	5.39	39.02	5.27
5	0.32	B	*140.41*	1.46	159.30	3.65	73.08	2.81
6	0.35	B	*149.85*	1.37	165.64	3.51	78.89	2.61
7	0.43	B	*173.42*	1.19	182.20	3.19	95.41	2.16
8	0.65	B	*245.85*	0.84	239.53	2.43	158.34	1.30
9	0.67	B	*253.56*	0.81	*246.11*	2.36	165.62	1.24
10	0.64	B	*242.17*	0.85	*236.42*	2.46	154.89	1.33
11	0.63	B	*241.96*	0.85	*236.24*	2.46	154.69	1.33
12	0.63	B	*241.96*	0.85	*236.24*	2.46	154.69	1.33
13	0.92	B	*337.68*	0.61	***321.68***	1.81	247.37	0.83
14	1.17	B	*424.10*	0.48	***403.65***	1.44	333.05	0.62
15	1.07	B	*390.20*	0.53	***371.17***	1.57	299.35	0.69
16	1.00	B	*366.23*	0.56	***348.43***	1.67	275.58	0.75

According to the standard [1], the maximum permissible normal stress σ_{dn} of the steel reinforced aluminium conductor equals 40% of the tensile strength σ_m of the aluminium wires, while the maximum permissible normal disastrous stress (for double ice loading) σ_{dnk} equals $0.8 \cdot \sigma_m$. For the AFL-6-240 conductor, $\sigma_{dn} = 119.91$ MPa and $\sigma_{dnk} = 239.82$ MPa. Table 1 shows (values in italic font) that after five years from the beginning of exploitation, the extension of the span caused by terrain deformation will exceed the allowable normal stress, and after nine years also the normal disastrous stress under the double ice loading. After 13 years (values in bold font) stress may exceed the tensile strength of the conductor, which may cause the line to break.

4. Factors mitigating the subsidence effects

The assumption has been made in the above calculations that the suspension points of the conductors are fixed and the supports are rigid constructions. In fact, elastic deformation of supports and their crossarms due to unbalanced longitudinal loads as well as the change in the angle of inclination of a suspension insulators reduce the stress and maximum sag [3].

The above factors can be interrelated. For example, the elastic deformation of the support depends not only on the parameters of this support and its type (tension support with strain insulators or intermediate support with suspension insulators), but also on the distribution of tension forces on both sides of support. The aforementioned interdependence of these factors makes it difficult to accurately (analytically) define their impact on decreasing conductors' stress. It is however possible to preset the range of compensation of span length changes. The limits of this range depend on the type of support forming the span under consideration. Thus, the lower limit of distance $\Delta a'_{min}$, by which the change in the span length will be reduced, can be determined by following principles:

- for a span bounded by two intermediate supports, $\Delta a'_{min}$ will be equal to half the sum of the inclination of the insulator chains on both supports and half of the sum of the elastic deformations of the crossarms:

$$\Delta a'_{min} = 0,5(\Delta a_{i1} + \Delta a_{i2}) + 0,5(\Delta a_{c1} + \Delta a_{c2}) \tag{6}$$

- for a span bounded by intermediate support from one side and tension support from the other, $\Delta a'_{min}$ will be equal to half the sum of the inclination of the suspension insulator, half of the elastic deformation of the tension support (at the height of the crossarm) and half of the elastic deformation of the crossarms of both supports:

$$\Delta a'_{min} = 0{,}5(\Delta a_i)_s + 0{,}5(\Delta a_t)_a + 0{,}5\,(\Delta a_c)_s + 0{,}5\,(\Delta a_c)_a \quad (7)$$

- for a span bounded by two tension supports, $\Delta a'_{min}$ will be equal to half the sum of elastic deformation of both supports and their crossarms:

$$\Delta a'_{min} = 0{,}5(\Delta a_{t1} + \Delta a_{t2}) + 0{,}5(\Delta a_{c1} + \Delta a_{c2}) \quad (8)$$

The upper limit of distance $\Delta a'_{max}$, by which the change in the span length will be reduced, can be determined as follows:

- for a span bounded by two intermediate supports:

$$\Delta a'_{max} = \Delta a_{i1} + \Delta a_{i2} + \Delta a_{c1} + \Delta a_{c2} \quad (9)$$

- for a span bounded by intermediate support and tension support:

$$\Delta a'_{max} = (\Delta a_i)_s + (\Delta a_c)_s + (\Delta a_t)_a + (\Delta a_c)_a \quad (10)$$

- for a span bounded by two tension supports:

$$\Delta a'_{max} = (\Delta a_{t1} + \Delta a_{t2}) + (\Delta a_{c1} + \Delta a_{c2}) \quad (11)$$

The following designations have been adopted in the above formulas:

Δa_t – elastic deformation of support's top,
Δa_c - elastic deformation of support's crossarm,
Δa_i – change of the span length due to inclination of suspension insulator,
s, a – indexes denoting intermediate support (s) and tension support (a).

Maximum values of above quantities for steel lattice towers can be adopted on the basis of the standard [5]:

- elastic deformation of intermediate support:

$$\Delta a_t = \frac{H_t}{70} \quad (12)$$

- elastic deformation of tension support:

$$\Delta a_t = \frac{H_t}{90} \quad (13)$$

- elastic deformation of support at the height of crossarm (index u denotes upper, l – lower crossarm)

$$\Delta a_{tu,l} = \Delta a_t \cdot k_{u,l} \quad (14)$$

where

$$k_{u,l} = \left| -\frac{3}{2} \cdot \left(\frac{H_{u,l}}{H_t}\right)^2 + \frac{1}{2} \cdot \left(\frac{H_{u,l}}{H_t}\right)^3 \right| \quad (15)$$

H_t, H_u, H_l – accordingly support's height, height at the upper crossarm mounting, height at the lower crossarm mounting,

- elastic deformation of crossarm

$\Delta a_c = \dfrac{q}{50}$ for intermediate supports

$\Delta a_c = \dfrac{q}{70}$ for tension supports

where q – length of a crossarm,

- maximum change of the span length due to inclination of suspension insulators depends on the length of insulator chain and maximum angle of inclination; for 110 kV lines $\Delta a_i = 0.63$ m.

Applying above principles, the values of elastic deformations of supports and the resulting length changes $\Delta a'_{min}$ and $\Delta a'_{max}$ for the exemplary span were calculated:

$$\Delta a'_{min} = 0.69 \text{ m} \qquad \Delta a'_{max} = 1.38 \text{ m}$$

Comparison of the mitigating effect of supports elasticity and insulators inclination with the expected change of the span length (second column of the table 1) shows, that the extension of the span due to terrain deformation can be partially or fully compensated. It should be emphasized that the abovementioned forecasting can only be preliminary. It is necessary to verify the results of calculations with the observation of the behavior of the line, especially in the initial stages of exploitation, when land deformations are not yet so large. Periodic inspections of power lines consist of, among others, on an approximate assessment of inclination of supports or measuring the angle of their inclination. However, it should be noted that the share of support inclination in the change of the span's length is relatively small (as a rule it does not exceed several percent) and the observation of only the angle of support inclination is not sufficient to obtain information about the increasing stress in the conductors. It is necessary to measure the span lengths and/or values of sag. Another advanced and more sophisticated techniques are available for monitoring the ground subsidence utilizing radars, GPS, satellites etc. [6]. The dates of performing control measurements should be coordinated with the course of underground mining activity.

5. Conclusions

The impact of subsidence on the operation of overhead power lines must be considered when planning underground exploitation, especially for high voltage lines, as they are important for the reliability of supply of mines. The results of calculations show, that in real lines operated in the area affected by mining damage, exceeding of the permissible stresses in conductors is possible. This situation can lead to breaking the lines or damage to the towers. Elastic deformations of towers and their crossbars as well as the inclination of suspension insulators can mitigate the effect of increasing the span's length, however analytical (quantitative) determination of this effect is difficult. Calculations results of stress and sag should be periodically verified by field observations to determine measures that should be taken to prevent damage to the line.

References

[1] PN-E-05100-1:1998 *Elektroenergetyczne linie napowietrzne – Projektowanie i budowa – Linie prądu przemiennego z przewodami roboczymi gołymi.*
[2] Bartodziej G et al 1990 *Sieci elektroenergetyczne w zakładach przemysłowych. Poradnik* (Warszawa: WNT)
[3] Kotlarski W and Kopia A 1978 *Linie napowietrzne na terenach robót górniczych. Wskazówki do projektowania (Gliwice:* Elektroprojekt)
[4] Knych T 2010 *Elektroenergetyczne przewody napowietrzne* (Kraków: Wydawnictwa AGH)
[5] PN-B-03205:1996 *Elektroenergetyczne linie napowietrzne – Stalowe konstrukcje wsporcze – Obliczenia statyczne i projektowanie.*
[6] Luo Q, Perissin D, Lin H, Zhang Y and Wang W 2014 *Subsidence Monitoring of Tianjin Suburbs by TerraSAR-X Persistent Scatterers Interferometry* Sel. Top. Appl. Earth Obs. Remote Sensing, IEEE J. **7** 1642–50

The possibilities for reducing mercury, arsenic and thallium emission from coal conversion processes

Piotr Burmistrz, Faustyna Wierońska, Marta Marczak and Dorota Makowska

AGH University of Science and Technology, Faculty of Energy and Fuels,
A. Mickiewicz Avenue 30, 30-059 Krakow, Poland

wieronsk@agh.edu.pl

Abstract. According to The European Parliament and The Council (Decision No. 1386/2013/UE), the emission of air pollutants during the last years has decreased, but still constitute a serious problem in many parts of Europe. Apart from sulfur oxide, nitrogen oxide and volatile organic compounds (VOCs), among air pollutants are also elements, which have a negative impact on environment and human health and life. During coal combustion and coking coal processes, a part of these elements are released into the atmosphere. For this reason, coal processing is one of the main sources of their emissions to the environment. Particular attention should be paid to the emission of such elements as mercury (Hg), arsenic (As), thallium (Tl) and their compounds, which are characterized by very high ecotoxicity. In this article the current standards and regulations on emissions of mercury, arsenic and thallium into the atmosphere for selected coal conversion processes as well as opportunities to reduce these emissions were presented.

1. Introduction

The amount of coal mined in Poland in 2016 amounted to over 130 million tons, 75% of which were coals combusted in the power industry, and more than 9% of coals were used to produce coke [1]. Coal, in addition to the main elements (C, H, O, N, S), contains the so-called ecotoxic elements. During the combustion and coking processes of coal, some of these elements are released into the atmosphere, causing that coal processing processes are one of the main sources of their emission to the environment. According to the European Parliament and the Council (1386/2013/EU), emissions into the air have declined in recent years, however, it is still a problem in many parts of Europe and requires taking measures to reduce emissions in particular: sulfur oxides, nitrogen oxides, volatile organic compounds (VOCs) and ecotoxic elements such as mercury (Hg), arsenic (As), thallium (Tl) and their compounds.

Mercury is a highly toxic heavy metal and the human body does not show physiological demands on it. This element, together with arsenic, has been classified by the US Environmental Protection Agency (US EPA) as dangerous air pollutants [2]. Hg and As show the most adverse environmental impact (Group I), while Tl is classified in Group III with slightly lower toxicity according to Swaine [3]. However, many sources compare toxicity of thallium to mercury or lead [4]. Mercury, arsenic and thallium are elements with a very high accumulation ratio [5]. Moreover, arsenic according to International Agency for Research on Cancer data has a proven carcinogenic effect [6]. The above elements are included in the list of dangerous substances set out in the Regulation of the Minister of Health of September 28th, 2005 in Poland [7].

The US Environmental Protection Agency adopted in 2012 a law introducing limitations in the emission of ecotoxic elements to the environment. These standards have been included both existing and newly built coal-fired power plants. Limits have been set for ecotoxic elements, including for mercury and arsenic [8]. Currently in Poland and in the European Union there are no specific limits regarding the emission of ecotoxic elements to the environment for combustion or coking coal. The only limits that have been introduced concern the certain industry sectors, including incineration and waste co-incineration plants in accordance with Directive 2010/75/EU and the Ordinance of the Minister of the Environment of the Republic of Poland of 4 November 2014 (table 1).

Table 1. Emission limit values of TE for plants incinerating or co-incinerating waste according to the Regulation of the Minister of the Environment of November 4, 2014.

Trace element (TE)/ sum of TEs	Content of TE in flue gases (mg/Nm^3)
Cd + Tl	0.05
Hg	0.05
Sb+As+Pb+Cr+Co+Cu+Mn+Ni+V	0.5

In addition, Poland is required to register quantities of ecotoxic elements (including mercury and arsenic) released into the air, water and soil on the basis of the arrangements approved during the Geneva Convention in 1979 on the long-range transboundary atmospheric pollutants transport [9] and in accordance with Regulation of the European Parliament and the Council 166/2006 [10]. The National Center for Emissions Management (KOBiZE) is responsible for developing the Polish report. According to data from KOBiZE and the European Environment Agency (EEA), Poland was the largest mercury emitter (18.6%) to the environment and one of the largest arsenic emitters (23.2%) from the EU-28 countries in 2015.

However, the legislative situation regarding the emission of ecotoxic elements to the environment is going to change, because in 2017 the European Commission adopted conclusions on the best available techniques BAT for large combustion plants [11]. The standards for CO_2, SO_2, NO_x and particulate matter PM emissions are going to be tightened, as well as mercury emission limits will been introduced. The mercury emission limit values varies from 1 to 10 μg/Nm^3, depending on the total thermal power delivered in the fuel and the type of fuel (hard coal, brown coal) (table 2). Furthermore, the BAT conclusions introduce the obligation to monitor emissions of ecotoxic elements, including arsenic and mercury to air (BAT 4) in accordance with EN 14385, EN 13211, EN 14884 standards, which involve the determination of these elements directly in the exhaust gas.

The standards regarding the emission of a total amount of Tl and Cd to the environment for co-incineration of waste with coal are also tightened (5-12 μg/Nm^3 for installations <300 MW, 5-6 μg/Nm^3 for installations ≥ 300 MW), biomass and peat (<5 μg/Nm^3). The above requirements and standards must be met by 2021. In the future, similar regulations are expected for other ecotoxic elements, including for arsenic and thallium. At present, the coke industry must meet the environmental requirements included in the IPPC Directive and BAT standards [12, 13].

Table 2. Levels of mercury emissions into the atmosphere from combustion process of hard coal and brown coal defined in BAT conclusions [11].

Combustion plant total rated thermal input (MW_{th})	BAT-AELs (μg/Nm^3) Yearly average or average of samples obtained during one year			
	New plant		Existing plant	
	Coal	Lignite	Coal	Lignite
< 300	< 1-3	< 1-5	< 1-9	< 1-10
≥ 300	< 1-2	< 1-4	< 1-4	< 1-7

2. Content and occurrence of mercury, arsenic and thallium in coal

The mercury content in coals is relatively low, from several dozen to several hundred μg/kg. The average mercury content in Polish subbituminous coal ranges from 25 to 300 μg/kg, while in Polish lignites varies from 100 to 450 μg/kg [14-17]. Mercury is present in coal both in a form associated with a mineral matter (mainly pyrite), an organic matter, e.g. in sulfuric connections and in the form of native mercury [18].

Arsenic in coal is mainly associated with pyrite. It can also occur as organic arsenous (As_{org}) and in combination with silicates [19]. The content of arsenic in coal is reported according to various literature sources within 0.5-80.0 mg/kg [20], 0.3-16.6 mg/kg [21] and 0.3-11.0 mg/kg [22]. There are also coals containing even up to 0.4% of As [23]. Its quantity in Polish bituminous coals and lignites is respectively 0-40 mg/kg [24] and 5-15 mg/kg [25].

Thallium in coal is mainly associated with pyrite. The content of this element in coal is in the range from 0.01-3 mg/kg [20, 26, 27]. A high concentration of thallium was found in lithotypes containing significant amounts of arsenic or sulfur (up to 26 mg/kg) [26]. In Polish hard coals and brown coals, the thallium content is 0.2-5.3 mg/kg and 0.2-2.4 mg/kg [28], respectively.

3. The possibilities for reducing mercury, arsenic and thallium emission from coal conversion processes

3.1. Coal combustion processes

The limitation of the emission of ecotoxic elements to the atmosphere from combustion processes can be divided into two groups of methods: primary (so-called pre-combustion) and secondary (post-combustion). The first group includes, among others, coal cleaning, descaling (dry separation), thermal preparation, chemical and biological methods, as well as selective coal mining [29 - 33]. Due to the connection of mercury, arsenic and thallium with coal mineral matter (mainly with sulfides) [16, 34, 35], coal cleaning processes achieve relatively high efficiency in removing these elements from coal. In the case of mercury, they allow to remove from 10 to 78% of this element [36 - 38]. The efficiency of arsenic removal in these processes ranges from 35% to 83% [39, 40], and from 41% to even 92% for thallium [41]. Research conducted on a pilot installation for dry separation (air concentrating table) also indicate the effectiveness of this method in removing mercury from coal [42]. Whereas, mercury associated with an organic matter can be successfully removed by thermal preparation. The best results in the removal of mercury from coal can be achieved using a hybrid method combining coal cleaning with thermal preparation [43].

The document Ref. Ares (2017) 1248230-09/03/2017 presents the BAT for reduction of mercury emissions, which can be used to reach the lower range of BAT standards for combustion of bituminous coal and lignite. These techniques are divided into two groups. The first includes the co-benefit methods, dedicated to removing other contaminants, but during which ecotoxic elements are removed by the way. Among these methods can be distinguished: selective catalytic reduction of nitrogen oxides (SCR), electrostatic precipitators (ESP), bag filters (FF), wet flue gas desulfurization (WFGD), dry/semi-dry flue gas desulfurization (D/SDFGD). The second group of methods includes special techniques of reducing mercury emission, in which one can distinguish among others: application of carbon sorbent (e.g. activated carbon or brominated activated carbon) injected into flue gases, use of halogenated additives in fuel or injection into the furnace.

In the gas formed after the combustion of coal, three speciation forms of mercury can be distinguished: oxidized mercury (Hg^{2+}), mercury associated with fly ash particles ($Hg_{(p)}$) and elemental mercury - metallic mercury (Hg^0). Due to the insolubility in water, durability and the ability to carry over long distances, the most undesirable form of mercury is Hg^0. The other two forms can be effectively removed in the wet desulfurization installation, as well as in dedusting devices such as ESP or FF [11, 31, 44] (table 3). According to Wang's research [45] the efficiency of the ESP in the mercury removal from flue gas varies from 6 to 46% and depends on the $Hg_{(p)}$ share in the flue gases and the efficiency of dust removal in the ESP. The use of SCR, through the catalytic oxidation of mercury Hg^0 to Hg^{2+}, can cause removal of 34% to 68% of the Hg^0 contained in the exhaust. On the other hand, the installation for removing sulfur oxides by the wet method can remove 15.3% - 80% of mercury from flue gas (including up to 97% of Hg^{2+} coming from the flue gas to FGD [45, 46]. The results of the research prove that the efficiency of flue gas cleaning from mercury depends largely on the applied exhaust gas treatment system and its effectiveness as well as on the occurrence of mercury in the exhaust gas, which is a consequence of the chemical composition of the coal burned and the type of the boiler [44].

Table 3. The effectiveness of mercury removal depending on the Air Pollution Control Devices (APCD).

Boiler	Coal (Hg content)	Air Pollution Control Devices	Effectiveness of Hg removal	Reference
PCB 200 MW	hard coal (233±12 μg/kg)	ESP+WFGD	68%	[45]
PCB 600 MW	hard coal (142±38 μg/kg)	ESP+WFGD	70%	
PCB 300 MW	anthracite (174±1 9 μg/kg)	ESP+WFGD	81%	
PCB 600 MW	brown coal (35±19 μg/kg)	ESP+WFGD	28%	
PCB 100 MW	hard coal (385±113μg/kg)	ESP+CFB-FGD+FF	66%	
PCB 165 MW	brown coal, (17±5 μg/kg)	SCR+ESP+WFGD	37%	
PCB 350 MW	hard coal (38.6 μg/kg) hard coal (33.7 μg/kg) brown coal (36.2 μg/kg)	PCB (low-emission burners) + SCR + combination of ESP/FF + WFGD	59% - 73%	[46]
PCB 370 MW	hard coal (66±9 μg/kg)	SNCR + ESP + WFGD	72% - 84%	[15]
PCB 225 MW	hard coal (100±15 μg/kg)	ESP + WFGD		
PCB 370 MW	brown coal (596±99 μg/kg)	ESP + WFGD		

PCB - pulverized-coal boiler; ESP - electrostatic precipitator; WFGD - wet flue gas desulfurization; CFB-FGD - circulating fluidized bed flue gas desulfurization; FF - fabric filter; SCR - selective catalytic reduction; SNCR - selective non-catalytic reduction.

Among the active methods, injection of powdered activated carbons and/or brominated activated carbons is the most commonly used. The effectiveness of mercury removal by means of the above method can range from 50% to over 90% (depending on the solution used) [47, 48]. However, with the increase in the mercury removal efficiency, the associated costs increase. The unit cost of removing mercury from flue gas using powdery activated carbons may amount to 40,000 - 90,000 $/kg [48].

Arsenic contained in coal goes into gas state in coal combustion process. The most common compounds of arsenic in gas are $As_{(g)}$, $As_2O_{3(g)}$. At high temperatures that prevail in boilers arsenic can also form solid As-Ca compounds ($Ca_3(AsO_4)_2$ and $Ca(AsO_2)_2$). Arsenic compounds are sorbent on fly ash particles when a flue gas temperature drops and are almost entirely removed from fly ash in an ESP or FF [19, 49]. The efficiency of cleaning the arsenic associated with ash separated on the electrostatic precipitator can reach up to 99.95%. According to Zhao [50] SCR is able to remove up to 29% of arsenic, which may be related to the condensation of arsenic compounds, in particular As_2O_3 in the micropores of the V_2O_5 catalytic converter [51, 52]. The efficiency of WFGD in purifying flue gas from arsenic can reach up to 71% according to Aunela-Tapola's research [53]. Depending on the treatment system used, the efficiency of removing As from gas can range from 98.85 - 99.9% [11, 54].

Thallium, same as arsenic, goes into the gas phase during coal combustion processes and then condenses on the fly ash particles. It is estimated that the amount of thallium released into the environment may be 700 - 2500 μg/m^3 [26]. On the other hand, according to of research [53], the use of an ESP in connection with the semi-dry desulfurization and bag filter installations results in lowering the Tl emission to the environment below 2.35 μg/m^3. Significantly lower emissions were obtained with the use of an ESP and wet desulfurization plant (table 4).

Table 4. Emissions of arsenic, mercury and thallium to the environment, depending on the on the APCD.

Boiler	Coal	Air Pollution Control Devices	TE emission As	Hg	Tl	Reference
PCB 660 MW	hard coal	SCR + ESP + WFGD + WESP	0.01 μg/m^3	-	-	[50]
PCB 160 MW	hard coal	PCB (low-emission burners) + ESP + S-DFGD + FF	<2.58 μg/m^3	1.88 μg/m^3	<2.35 μg/m^3	[53]
PCB 113 MW	hard coal	PCB (low-emission burners) + ESP + S-DFGD + FF	<2.03 μg/m^3	<0.11 μg/m^3	<1.98 μg/m^3	
PCB 350 MW	hard coal	PCC (low-emission burners) + SCR + combination of ESP/FF + WFGD	0.11 μg/m^3	-	-	[55]
PCB 350 MW	brown coal	PCC (low-emission burners) + SCR + combination of ESP/FF + WFGD	0.20 μg/m^3 0.25 μg/m^3	-	-	
PCB	hard coal	ESP + WFGD	2.3 μg/m^3	8.8 ng/m^3	0.1 μg/m^3	[56]
PCB 660 MW	-	SCR + ESP + WFGD + WESP	0.01 μg/m^3 0.05 μg/m^3	-	-	[54]

PCB - pulverized-coal boiler; ESP - electrostatic precipitator; WFGD - wet flue gas desulfurization; FF - fabric filter; SCR - selective catalytic reduction; WESP - wet electrostatic precipitator; S-DFGD - semi-dry flue gas desulfurization.

3.2. Coal coking processes

The reduction of mercury emissions from coking processes can be obtained by using coal enrichment processes, which will allow the removal of between 21% - 84% of mercury contained in the fuel before the coking process [57]. In the coal coking process, the ecotoxic elements contained in coal pass into coke: 6.0% - 23.4% for Hg, 51.6% - 97.21% for As and 31.3% - 99% for Tl. Such broad ranges may be caused by various forms of occurrence of the above elements in coal [13]. During coal coking, the ecotoxic elements may be emitted to the environment during coke oven cell filling, coke pushing out from the chamber, during the process of quenching coke and burning coke oven gas in heating channels of the battery. The content of arsenic in the gas collected during charging of the coking chamber, pushing out coke and burning coke oven gas were respectively: 1.7 μg/m^3, 76.8 μg/m^3 and 2.3 μg/m^3 [58]. According to the requirements presented in the BAT conclusions adopted for coke installations, the limitation of the above fugitive emission of dust and thus ecotoxic elements to the environment can be obtained by filling coking chambers using low-emission filling systems and by using special exhaust hoods during coke pushing out from chamber. It is also recommended to use of dry coke quenching with the removal of coking dust using dedusting devices [59].

4. Conclusion

The paper reviews the regulations and standards for mercury, arsenic and thallium emissions to the environment from coal combustion and coking processes as well as the possibilities of reducing their emissions to the environment.

- In the BAT conclusions adopted in 2017, in addition to the tightening of the CO_2, SO_2, NO_x and PM emission standards, mercury emission limits have been introduced for new and existing coal combustion plants, which will apply in the EU power industry from 2021.
- In order to meet mercury emission standards for combustion processes, it will be important to use both primary and secondary methods. In the case of insufficient effectiveness of passive secondary methods (installation for catalytic selective reduction of NO_x and SO_2, PM removal in ESP or FF) additional methods will be required, e.g. the most commonly used dusty sorbent injection.
- Due to the comparable toxicity of arsenic and thallium to mercury and the European Union's environmental policy conducted in the future, standards for other ecotoxic elements may be expected to be introduced.
- Currently, there are no regulations regarding the emission of ecotoxic elements to air from coal coking processes.
- The emission of arsenic during the coke ejection process can amount to as much as 76.8 μg/m^3. This is a value much higher than the arsenic emission to the environment from combustion processes.

Acknowledgment
Work carried out in the framework of the statutory AGH No. 11.11.210.373

References

[1] ARE 2016 Sytuacja Energetyczna w Polsce - Krajowy Bilans Energii
[2] US EPA 1998 A study of Hazardous Air Pollutant emissions from electric utility steam generating units: Final Report to Congress. EPA-453/R-98-004a US EPA Office of Air Quality Planning and Standards (US Government Printing Office, Washington)
[3] Swaine D J 2000 Why trace elements are important *Fuel Processing Technology* **65–66** pp. 21–33
[4] John Peter A and Viraraghavan T 2005 Thallium: a review of public health and environmental concerns *Environment International* **31** pp. 493–501
[5] Srogi K 2007 Pierwiastki śladowe w węglu *Wiadomości Górnicze* **2** pp. 87–96
[6] International Agency for Research on Cancer Agents Classified by the IARC Monographs **1–117**
[7] Rozporządzenie Ministra Zdrowia z dnia 28 września 2005 r. w sprawie wykazu substancji niebezpiecznych wraz z ich klasyfikacją i oznakowaniem (Dz.U.05.201.1674)
[8] USEPA, Federal Register **77** no. 32, part II, National Emission Standards for Hazardous Air Pollutants From Coal- and Oil-Fired Electric Utility Steam Generating Units and Standards of Performance for Fossil-Fuel-Fired Electric Utility, Industrial-Commercial- Institutional, and Small Industrial-Commercial-Institutional Steam Generating Units; Final Rule
[9] Convention on Long-range Transboundary Air Pollution 1979 Genewa
[10] REGULATION (EC) No 166/2006 OF THE EUROPEAN PARLIAMENT AND OF THE COUNCIL of 18 January 2006 concerning the establishment of a European Pollutant Release and Transfer Register and amending Council Directives 91/689/EEC and 96/61/EC
[11] Best Available Techniques (BAT) Reference Document for Large Combustion Plants, Final Draft Joint Research Centre June 2016
[12] Machowska H 2011 Przemysł koksowniczy w aspekcie ochrony środowiska *Proceedings of EC Opole* **5**
[13] Konieczyński J, Zajusz-Zubek E and Jabłońska M 2012 The release of Trace Elements in the

process of Coal Coking *The Scientific World Journal* **2012**

[14] Burmistrz P and Kogut K 2016 Mercury in Bituminous Coal Combusted in Polish Power Plants *Arch. Min. Sci.* **61** issue 3 pp. 473–88

[15] Burmistrz P, Kogut K, Marczak M and Zwoździak J 2016 Lignites and subbituminous coals combustion in Polish power plants as a source of anthropogenic mercury emission *Fuel Process. Technol.* **152** pp. 250–58

[16] Dziok T, Strugała A, Rozwadowski A and Macherzyński M 2015 Studies of the correlation between mercury content and the content of various forms of sulfur in Polish hard coals *Fuel* **159** pp. 206–13

[17] Wichliński M, Kobyłecki R and Bis Z 2013 The investigation of mercury contents in Polish coal samples *Arch. Environ. Protec.* **39** issue 2 pp. 141–50

[18] Yudovich Y E and Ketris M P 2005 Mercury in coal: a review – Part 1. *Geochemistry International Journal of Coal Geology* **62** pp. 107–34

[19] Yudovich Ya E and Ketris M P 2005 Arsenic in coal: a review *International Journal of Coal Geology* **61** pp. 141– 96

[20] Swaine D J 1990 *Trace Elements in Coal* (London: Butterworths) p. 292

[21] Saha D, Chakravarty S, Shome D, Basariya M R, Kumari A, Kundu A K, Chatterjee D, Adhikari J and Chatterjee D 2016 Distribution and affinity of trace elements in Samaleswari coal, Eastern India *Fuel* **181** pp. 376–88

[22] Dai S, Ren D, Tang Y, Yue M and Hao L 2005 Concentration and distribution of elements in Late Permian coals from western Guizhou Province, China International *Journal of Coal Geology* **61** pp. 119–37

[23] Karayigit A I, Spears D A and Booth C A 2000 Antimony and arsenic anomalies in the coal seams from the Gokler coalfield, Gediz, Turkey *International Journal of Coal Geology* **44** pp. 1–17

[24] Widawska-Kuśmierska J 1981 Występowanie pierwiastków śladowych w polskich węglach kamiennych *Przegląd Górniczy* **7-8** pp. 455–9

[25] Wagnar M 2001 Oznaczenie pierwiastków toksycznych i szkodliwych w węglu i jego popiele *Eksploatacja węgla brunatnego jako metoda ograniczania szkodliwego oddziaływania na środowisko pierwiastków obecnych w węglu i w produktach jego spalania (na przykładzie KWB Bełchatów)* ed M Stryszewski (Kraków: Katedra Górnictwa Odkrywkowego Wydział Górniczy AGH)

[26] López Antón M A, Spears D A, Díaz Somoano M and Martínez Tarazona M R 2013 Thallium in coal: Analysis and environmental implications *Fuel* **105** pp.13–8

[27] Dai S, Ren D, Tang Y, Yue M and Hao L. 2005 Concentration and distribution of elements in Late Permian coals from western Guizhou Province *China International Journal of Coal Geology* **61** pp. 119–37

[28] Bojakowska I and Paulo A 2013 Thallium in mineral resources extracted in Poland *E3S Web of Conferences* **1** 14006

[29] Burmistrz P, Dziok T, Kogut K and Makowska D 2014 Methods of Mercury Content Reduction in Coal *Mercury As a Coal Combustion Pollutant* ed J Gołaś and A Strugała (Warszawa: Oficyna Drukarska – Jacek Chmielewski)

[30] Kurus K and Białecka B 2015 Możliwości i ograniczenia redukcji ładunku rtęci na etapie produkcji węgla kamiennego w Polsce *Systemy wspomagania w Inżynierii Produkcji* **3** pp. 90-8

[31] Wichliński M, Kobyłecki R and Bis Z 2012 Przegląd metod ograniczenia emisji rtęci w elektrowniach podczas spalania paliw stałych *Polityka Energetyczna* **15** pp. 151–60

[32] Stryszewski M 2001 Metoda eksploatacji selektywnej w odniesieniu do koncentracji i rozmieszczenia pierwiastków toksycznych i promieniotwórczych w węglu brunatnym *Eksploatacja selektywna węgla brunatnego jako metoda ograniczania szkodliwego oddziaływania na środowisko pierwiastków obecnych w węglu i w produktach jego spalania*

(na przykładzie KWB Bełchatów) (Kraków: Katedra Górnictwa Odkrywkowego Wydział Górniczy Akademia Górniczo-Hutnicza)

[33] Makowska D, Bytnar K, Dziok T and Rozwadowska T 2014 Wpływ procesu wzbogacania na zawartość niektórych metali ciężkich w polskich węglach kamiennych *Przemysł Chemiczny* **93** pp. 2048–53

[34] Hower J C, Campbell J L, Teesdale W J, Nejedly Z and Robertson J D 2008 Scanning proton microprobe analysis of mercury and other trace elements in Fe-sulfides from a Kentucky coal *International Journal of Coal Geology* **75** pp. 88–92

[35] Mastalerz M and Drobniak A 2007 Arsenic, cadmium, lead, and zinc in the Danville and Springfield coal members (Pennsylvanian) from Indiana *International Journal of Coal Geology* **71** pp. 37–53

[36] Streets D, Hao J, Wang S and Wu Y 2009 Mercury emissions from coal combustion in China *Mercury Fate and Transport in the Global Atmosphere* ed R Mason and N Pirrone (Boston: Springer MA) pp. 51–65

[37] Dziok T, Strugała A, Rozwadowski A, Górecki J and Ziomber S 2014 Zmiany zawartości rtęci w węglu kamiennym w procesie jego wzbogacania *Polityka Energetyczna* **17** (4) pp. 277–88

[38] Zajusz-Zubek E and Konieczyński J 2014 Coal cleaning versus the reduction of mercury and other trace elements emission from coal combustion processes *Archives of Environmental Protection* **40** issue 1 pp. 115–27

[39] Zhu C, Tian H, Cheng K, Liu K, Wang K, Hua S, Gao J and Zhou J 2016 Potentials of whole process control of heavy metals emissions from coal-fired power plants in China *Journal of Clear Production* **114** pp. 343–51

[40] Akers D and Dospoy R 1994 Role of coal cleaning in control of air toxics *Fuel Processing Technology* **39** pp. 73-86

[41] Makowska D, Strugała A, Świątek K and Wierońska F 2017 Possibility of decreasing the emission of thallium from hard coal conversion processes through coal cleaning *Book of abstracts SEED 2017 International Conference on the Sustainable Energy and Environmental Development* Kraków p. 216

[42] Baic I, Blaschke W, Dziok T, Strugała A and Sobko W 2017 Badanie podatności węgli energetycznych na zmniejszenie zawartości rtęci na etapie pre-combustion *Zeszyty Naukowe Instytutu GSMiE PAN* **98** pp. 103–14

[43] Dziok T, Strugała A, Chmielniak T, Baic I and Blaschke W 2017 Koncepcja hybrydowego procesu usuwania rtęci z węgla kamiennego *Zeszyty Naukowe Instytutu GSMiE PAN* **98** pp. 125–35

[44] Bujny M, Burmistrz P, Gruszka S, Janicki W, Kogut K and Strugała A 2012 Instalacja demonstracyjna do monitorowania i redukcji emisji rtęci ze spalania węgla kamiennego w kotłach pyłowych *Polityka Energetyczna* **15** pp. 161–73

[45] Wang S X, Zhang L, Li G H, Wu Y, Hao J M, Pirrone N, Sprovieri F and Ancora M P 2010 Mercury emission and speciation of coal-fired power plants in China *Atmospheric Chemistry and Physics* **10** issue 3 pp. 1183–92

[46] Zhao S, Duan Y, Chen L, Li Y, Yao T, Liu S, Liu M and Lu J 2017 Study on emission of hazardous trace elements in a 350 MW coal-fired power plant. Part 1. Mercury *Environmental Pollution* **229** pp. 863–70

[47] Liberti L, Notarnicola M, Amicarelli V, Campanaro V, Roethel F and Swanson L 1998 Mercury removal with powdered activated carbon from flue gases at the Coriano municipal solid waste incineration plant *Waste Management & Research* **16** issue 2 pp. 183–89

[48] Burmistrz P, Czepirski L, Kogut K and Strugała A 2014 Removing mercury from flue gases. A demo plant based on injecting dusty sorbents *Przemysł Chemiczny* **93/12** pp. 2014–9

[49] Shen F, Liu J, Zhang Z and Dai J 2015 On-line analysis and kinetic behavior of arsenic release during coal combustion and pyrolysis *Environ. Sci. Technol.* **49** (22) pp. 13716–23

[50] Zhao S, Duan Y, Tan H, Liu M, Wang X, Wu L, Wang C, Lv J, Yao T, She M and Tang H

2016 Migration and Emission Characteristics of Trace Elements in a 660MW Coal-fired Power Plant of China *Energy and Fuels* **30** issue 7 pp. 5937–44

[51] Li J, Pengi Y, Chang H, Li X, Crittenden J C and Hao J 2016 Chemical poison and regeneration of SCR catalysts for NOx removal from stationary sources *Frontiers of Environmental Science & Engineering* **10** issue 3 pp. 413–27

[52] Li X, Li J, Peng Y, Si W, He X and Hao J 2015 Regeneration of Commercial SCR Catalysts: Probing the Existing Forms of Arsenic Oxide *Environmental Science and Technolgy* **49** pp. 9971–8

[53] Aunela-Tapola L, Hatanpaa E, Hoffren H, Laitinen T, Larjava K, Rasila P and Tolvanen M 1998 A study of trace element behaviour in two modern coal-fired power plants II. Trace element balances in two plants equipped with semi-dry flue gas desulphurisation facilities *Fuel Processing Technology* **55** pp. 13–34

[54] Zhao S, Duan Y, Wang C, Liu M, Lu J, Tan H, Wang X and Wu L 2017 Migration Behavior of Trace Elements at a Coal-Fired Power Plant with Different Boiler Loads *Energy and Fuels* **31** pp. 747–54

[55] Zhao S, Duan Y, Chen L, Li Y, Yao T, Liu S, Liu M and Lu J 2017 Study on emission of hazardous trace elements in a 350MWcoal-fired power plant. Part 2. arsenic, chromium, barium, manganese, lead *Environmental Pollution* **226** pp. 404–11

[56] Córdoba P, Ochoa-Gonzalez R, Font O, Izquierdo M, Querol X, Leiva C, López-Antón M A, Díaz-Somoano M, Martinez-Tarazona M R Fernandez C and Tomás A 2012 Partitioning of trace inorganic elements in a coal-fired power plant equipped with a wet Flue Gas Desulphurisation system *Fuel* **92** pp. 145–57

[57] Dziok T 2016 Badania zmiany zawartości rtęci na drodze przeróbki mechanicznej i wstępnej preparacji termicznej węgli kamiennych Doctoral Thesis Kraków

[58] Mu L, Peng L, Liu X, Bai H, Song C, Wang Y and Li Z 2012 Emission characteristics of heavy metals and their behavior during coking processes *Environmental Science and Technology* **46** pp. 6425–30

[59] COMMISSION IMPLEMENTING DECISION of 28 February 2012 establishing the best available techniques (BAT) conclusions under Directive 2010/75/EU of the European Parliament and of the Council on industrial emissions for iron and steel production

Environmental problems on the areas of redeveloped hard coal deposits

Henryk Kleta and Franciszek Plewa

Faculty of Mining and Geology, Silesian University of Technology, ul. Akademicka 2, 44-100 Gliwice, Poland

henryk.kleta@polsl.pl

Abstract. The issue of development of deposits in the areas of liquidated mines has been the subject of business activities in Poland for a number of years. This state is undoubtedly influenced by the fact that the Polish energy industry is based on coal and the needs of the steel industry for coking coal. The article presents the basic environmental problems that should be taken under consideration in the areas of re-utilized hard coal deposits. On the basis of selected examples the environmental conditions necessary to be taken into account when planning the re-development of hard coal deposits have been discussed.

1. Introduction

Hard coal deposits are one of the most important components of the national wealth. The volume as well as the accessibility of the deposits determine the national energy security. The hard coal deposits in Poland are found in three basins. Coal is mined in two of them, namely in the Upper-Silesian Coal Basin and the Lublin Coal Basin. Currently, the documented recoverable reserves of hard coal as of 31.12.2016 amount to 58 578 million tons. These include nearly 71.56% of power coal and over 27.09% of coking coal [1]. The environmental conditions are currently the most significant factor when it comes to the possibility of development of hard coal deposits. In post-mining areas that are affected by former mining activity, the changes which cause degradation of the ground surface include especially [2]:

- transformation of surface due to underground and surface mining exploitation,
- deposition of mining wastes and the space required for that purpose,
- contamination of soil with industrial emissions,
- transformation and erosion of soil in agricultural and forest areas.

The greatest hazard for the surface area is posed by mining exploitation conducted at low depths with a caving system without backfilling. The moment of formation of sink holes is practically impossible to forecast which causes a hazard to the surface and all the buildings at the affected area [3]. The collapse of voids resulting from exploitation and maintaining stability in the rock mass for a longer period of time may be caused by another mining exploitation as well as when load is exerted on the surface of the void or during rock-bursts and vibrations of the rock mass and the surface area [4].

2. "Morcinek 1" hard coal deposit

2.1. Short characteristics of the liquidated KWK "Morcinek" mine

The "Morcinek" hard coal mine was liquidated in the years 1998-2002. It was one of the youngest mines in Poland as it was constructed in the years 1978-1993. The exploitation was officially commenced in 1986. On October 30th 1998, the exploitation was terminated and the liquidation process was initiated. The liquidation was completed in 2001 [5]. For the KWK "Morcinek" mine and for the protection of the border zone between Poland and Czech Republic, protective pillars were established, which, along with the pillars for the protection of the main headings, took up a large area and significantly limited the development of the mine. Between the depths of approx. 600 m and 1500 m, the presence of 37 coal seams was found within the range from seam 354 to the seam 510. For the depth of 1100 m, that is the depth of the bottom level of the former "Morcinek" mine, 21 recoverable coal seams have been identified in the range between seam 360/1 and 418/2. Generally, the calorific value of coal increases along with the depth, reaching classes from 34.2 in the same seam (404/1, 4,4/1-2) to 37.2. During reconnaissance works, also natural coking of a coal seam was noted to proceed after the coal of the class 37.2. In lower seams, 42-type coking coal is found.

During the construction of the KWK "Morcinek" coal mine, in the Carboniferous formations water inflow occurred in sections of the roof, especially in sandstones beneath Dębowiec conglomerates. In regions that are more distant from the Dębowiec formations and at greater depths, the rock mass was practically free of water and it did not accumulate in parts with more cracks and in locations of faults transitions. The slight inflows that were noted were mostly relic waters and quickly faded. The main inflow was always constituted by waters from the overburden. The inflow of water to the mine has been presented in table 1.

Table 1. Inflow of waters (in January) into the "Morcinek" coal mine during mining operations in years 1986-1998 in m^3/min [6].

Year	1986	1987	1988	1989	1990	1991	1992	1993	1994	1995	1996	1997	1998
Total	0,56	0,95	2,15	2,98	2,48	2,08	1,52	1,43	1,70	1,73	1,87	1,29	1,71
Shafts	0,45	0,44	0,38	0,31	0,28	0,24	0,24	0,18	0,17	0,25	0,19	0,26	0,22
Overburden	0,45	0,76	1,96	2,39	2,22	1,69	1,30	1,11	1,20	1,44	1,54	1,20	1,60
Carboniferous	0,11	0,19	0,19	0,58	0,26	0,39	0,22	0,32	0,50	0,29	0,33	0,09	0,11

2.2. The "Morcinek 1" hard coal deposit and the characteristics of the terrain in its area

The "Morcinek 1" hard coal deposit is located in the Silesian voivodeship (figure 1). It lies within the outline of Zebrzydowice (Kaczyce) and Hażlach (Pogwizdów) communes.

The landscape of the area above the "Morcinek 1" hard coal deposit is highly diversified, exhibiting numerous elevations, valleys and steep slopes by streams. The area in concern is cut by the Olza river – the right-side tributary of Odra, as well as the Piotrówka river. Olza flows in a vast, flat valley along the western boundary of the area in concern. The watercourses which serve the drainage of the area are the right-sided tributaries of Olza and left-sided tributaries of Piotrówka. These also include the Kaczycki (Kaczor) stream and the unnamed streams constituting indirect tributaries of Olza. Moreover, surface water bodies located along the streams are present in the area in concern. The surface of the area of the "Morcinek 1" deposit is also developed with scattered residential and service buildings as well as the Kaczyce, Brzezówka and Pogwizdów villages with cluster development. The respective villages and developments are connected by a network of local roadways. Also the Zebrzydowice-Cieszyn railway runs through the area in concern. Moreover, drinking water pipelines, gas lines, communication lines and power lines are found in the area.

2.3. General characteristics of the effects of the exploitation conducted by the liquidated KWK "Morcinek" coal mine

The twelve years of mining activity of the liquidated "Morcinek" mine have had an impact on the transformation of the surface, which was subject to both intentional transformations accompanying the deposition of wastes and unintentional effects accompanying the underground mining activity, causing continuous and discontinuous surface deformations. The morphology of the area exhibit forms that are both concave (subsidence troughs) and convex (an embarked and filled sedimentation basin and a barren rock dump).

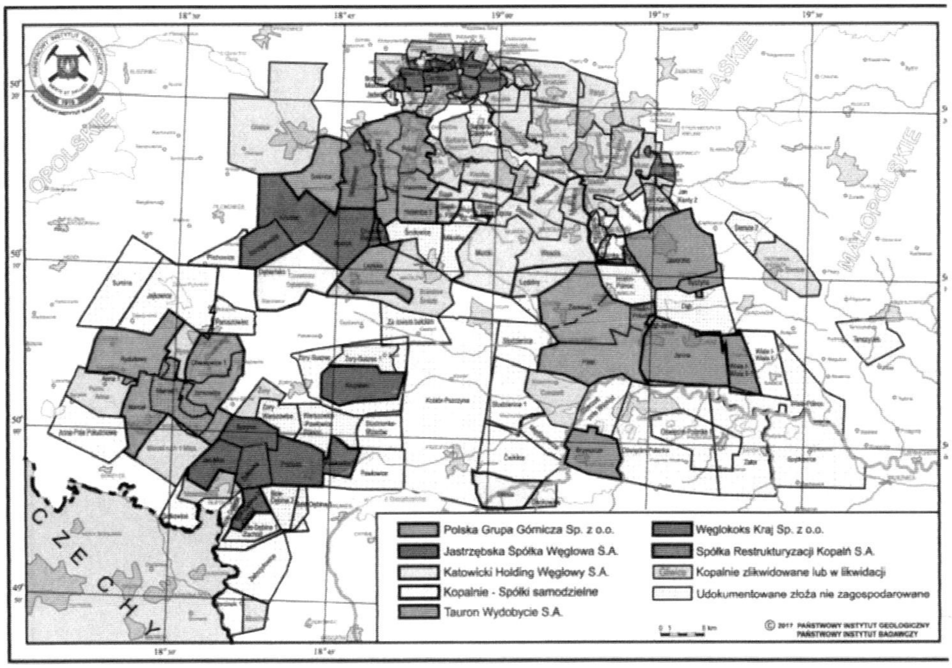

Figure 1. Areas of coal deposits GZW [www.pgi.gov.pl].

The effects of the mining exploitation in the form of subsidence were found at an area of approximately 5.6 km^2, encompassing developed areas, including mostly the scattered agricultural development in the Kaczyce and Pogwizdów villages. The maximal levels of subsidence occurred in two regions [6]. In the first region the subsidence reached approximately 8 m, while in the second region, the maximal values reached approx. 2.5 m. The subsidence has not caused significant hydrogeological changes (overflow) on the surface.

The mining exploitation formerly conducted in the liquidated "Morcinek" coal mine has not caused discontinuous deformations. The deformations that have emerged have not disturbed the hydrological ratios in the former "Morcinek" mine exploitation area. Also, no significant changes of hydrographical and hydrogeological conditions such as overflows, flooding, changes in stream gradients and levels were noted.

As a result of the drainage of the liquidated KWK "Morcinek" coal mine, a cone of depression was formed in the hydraulically connected complex of the waterlogged strata in the area of the Dębowiec formations and the water-bearing Carboniferous formations. The cone, which was located in the region of the main shafts I, II and II (formerly the "Morcinek" mine), reached the depth of approximately 1000 m and reached the radius of approx. 3300 m, stretching to the Dębowiec formations.

The geological documentation states that 14.46 million m^3 of water was pumped out during the existence of the "Morcinek" coal mine. After the pumping has ended, the mine would be gradually flooded. It was forecasted that after 4 years, the entire level 950 would be flooded up to the floor of the headings at the 800 m level (-550 m a.s.l.). After 9 more years, the water was supposed to fill the headings of the mine up to the floor of the 650 level and continue to fill the cone in the Dębowiec formations [5, 7].

The characteristics of the water filling the goafs of the former KWK "Morcinek" coal mine are highly variable depending on the location. In the region near the outcrop, beneath the Dębowiec formations in the western part, the mineralization of water is the lowest due to the highest infiltration of external water. During the construction of the Kaczyce shafts, the mineralization of water in the

IOP Conf. Series: Earth and Environmental Science **174** (2018) 012004 doi:10.1088/1755-1315/174/1/012004

Dębowiec formations and at the level of the roof of the Carboniferous formations was from 7 to 15 g/dm^3. Along with the depth, the salinity of water increased, reaching the mineralization from several dozen up to over 100 g/dm^3 and – in some locations – relic water occurred with mineralization reaching up to 200 g/dm^3.

Until 2013, natural flooding of the goafs proceeded and methane was exploited from the anthropogenic deposit that has formed in the goafs, based on the license to conduct a reconnaissance of methane deposits from coal seams in the "Kaczyce 1" mining area, granted by the Minister of Environment in the year 2000.

In case of establishing a new mine for the "Morcinek 1" deposit, the main hazard will be constituted by the methane from the gas reservoir formed in the goafs of the liquidated KWK "Morcinek" coal mine. Due to the chemical composition, the water in the Carboniferous formations will exhibit magnesium aggressiveness and may be acidic and contain aggressive CO_2. This will be accompanied by high outflow of methane. Moreover, the temperature of the Carboniferous water may be high and the natural water outflowing from the goafs will be warm. In case of an outbreak of water from the "Morcinek" coal mine, it should be expected that the temperature of water may reach 40–50°C.

3. "Brzezinka 3" deposit in Mysłowice

3.1. General characteristics of the effects of the exploitation conducted by the liquidated KWK "Morcinek" coal mine

The "Brzezinka 3" coal deposit is located within the administrative territory of the Mysłowice city. The area of the deposit is 7.27 km^2.

The "Brzezinka 3" deposit (figure 2) is adjacent to the developed "Mysłowice," "Wesoła" and "Brzezinka 1" deposits, the latter being documented by Tauron Wydobycie S.A. – Sobieski Mining Plant and the "Niwka-Modrzejów" and "Jan Kanty" deposits formerly exploited by liquidated mines.

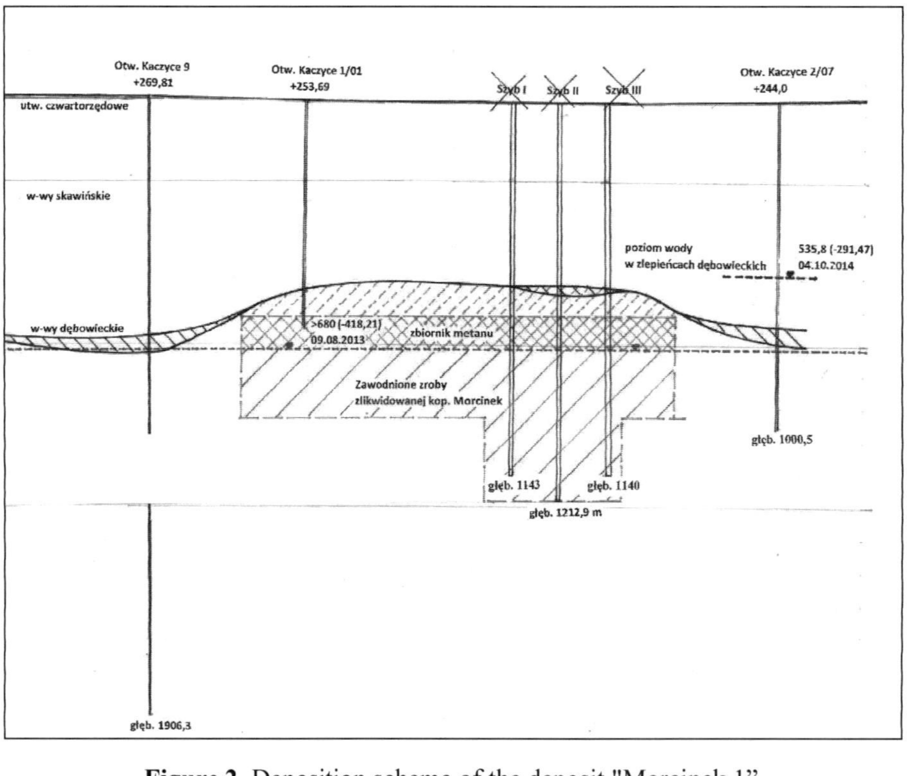

Figure 2. Deposition scheme of the deposit "Morcinek 1".

The deposit is documented and classified in the B assessment category from the roof of the saddle beds up to the floor of the 510 seam, that is, the depth of 1050 m. The reserves of hard coal in "Brzezinka 3" are deposited in the 510 seam of the saddle beds. The coal is of 31 and 32 types.

The 510 seam in the "Brzezinka 3" deposit is of a considerable thickness and is found at depths from approx. 700 m in the northern part of the deposit up to 1050 m at the south [8]. Thickness lower than 4.0 m (1.1 m), has only been noted in the Brzezinka 10 borehole located in the south, right behind the boundary of the deposit. The 510 seam was formerly exploited in adjacent deposits: „Niwka-Modrzejów," „Mysłowice" and „Wesoła".

3.2. Hazards posed to the surface by old goafs resulting from past mining exploitation

The beginnings of mining in the area outlined by the "Brzezinka 3" deposit date back to the first half of the 19th century. In the deposit in concern, shallow 215, 216 and 304 seams were exploited by the former "Przemsza" mine and the resulting vast goafs at the depths of 150 ÷ 200 m are probably flooded [8].

Until 1931, over a dozen of mining plants existed in the area in concern, including i.a.: "Theodor," "Louise," "Leopoldine," "Hans," "Wanda," "Nowa Przemsza," "Bartelmus," "Carlssegen," "Friedrichsgluck," "Maximilian," "Josepha," all of which conducted mining activity. The deposit was mostly exploited using a shortwall system with caving and a system of headings. In outcrop areas, the coal seams were made available using drifts and – rarely – the deposit was exploited using open pits.

In the years 1971-1982, the "Jan Kanty" mine exploited the 304/2 seam over the northern part of the "Brzezinka 3" deposit using open-end mining with caving and a longwall system with hydraulic supports and caving (to a lesser extent).

Based on the data regarding the shallow exploitation within the area of the "Brzezinka 3" deposit and the cases of discontinuous deformations occurring formerly at that area, it may be concluded that in case of commencing an exploitation of that deposit, significant hazard may be posed to the surface area.

The case of the M. Rej housing estate in Mysłowice – Brzezinka may serve as an example illustrating the subsidence hazard in that area. The M. Rej housing area is constituted by free-standing or semi-detached residential houses, mostly 2-floor, with full basements and small ground-plan dimensions. The 301 seam in the area of the M. Rej estate was exploited by the "Nowa Przemsza" coal mine. The exploitation was conducted in the years 1851 – 1885 and 1929 – 1930 using a shortwall system with caving at a mean depth of approx. 37 m. The thickness of exploitation was from 3 m to 4 m. As it has been shown by boreholes, the headings have not fully collapsed and thus pose a hazard to the surface area [4]. In the years 1880 – 1912, seam 304 (Luiza) was exploited beneath the seam 301 and the exploitation was conducted using a shortwall system with caving with exploitation thickness of approx. 4.0 m. The depth of the exploitation was approx. 170 m. Due to the thick benches of sandstone present between the seams 301 and 304, the exploitation of the 304 seam did not cause the reactivation of the goafs of the 301 seam and the collapse of the voids that are present therein.

The documentation of the phenomena occurring at the area of the M. Rej estate in Mysłowice – Brzezinka since 1977 indicates that the area of that estate was subject to typical shallow exploitation, where discontinuous deformations occur. Below, the most significant discontinuous deformations which have occurred at the area of the estate have been listed.

The most hazardous effects were caused by a sink hole in the form of an elliptical cone with the dimensions of 3.5 x 5.0 m and the depth of 4.0 m, that formed on May 3rd 1977 in the area of Różana street. Four detached family houses and a transformer station were in the range of the sink hole. A breakage of a water pipeline and outflow of water was noted in the sinkhole. The sinkhole was backfilled, however, due to the continued subsidence, the building No. 3 was damaged beyond repair. The liquidation of the sinkhole was completed on 14.07.1977.

On August 2nd. 1977, 3 sinkholes were noted in the area of Malinowa street. On August 3rd 1977, a sinkhole was noted by Różana street, in a garden, next to a cesspit. The dimensions of the sinkhole were 1.0 m x 0.5 m and its depth was 1.0 m.

Also on August 3rd 1977, a sinkhole was noted in the area of Jaśminowa street, with a volume of approx. 6 m^3. In 1979, also in the vicinity of a building at Jaśminowa street, a sinkhole was formed with a diameter of 2.0 m and a depth of 4.5 m. In 1982, the sinkhole renewed reaching the depth of 4.5 m. In 1979, a sinkhole appeared at Jaśminowa street with a diameter of approx. 2.0 m and a depth of 1.0 m. In June 1985 a sinkhole formed in the area of a residential building at M. Rej street. This sinkhole, with a diameter of 3.0 m and the depth of 0.8 m, occurred in the area of the "Stefania" shaft reaching the 301 seam. The sinkhole was backfilled and in September and October 1985, 5 boreholes were made in that area, while the boreholes No. 1 and 2 were filled with pressure-pumped fly-ash.

While analyzing the geological and mining conditions and the sinkholes that formed to that moment, it should be concluded that the hazards caused by discontinuous deformations in the M. Rej residential area have been affected by several factors, especially:

- the existence of a former shallow exploitation conducted using caving system in the 301 seam at the depth of approx. 33 m to 37 m,
- the existence of voids in the goafs of the 301 seams and over the 301 seam at varying depths, found by means of boreholes, electro-resistivity and gravimetric measurements,
- the lack of an insulating layer in the roof of the Carboniferous formations, causing the infiltration of precipitation water and the water from the damaged water pipelines and soakaways into the rock mass,
- shocks caused by the traffic within the residential area and – especially – outside of the area (express road).

It is commonly assumed that the process of void formation proceeds according to a certain model based on the so-called pressure arch theory [2]. In case of wider headings, the void reaches the surface and forms a sinkhole. At lower widths of headings, the void should be bound within the rock mass area.

In such case, for the sinkhole to form, additional circumstances must occur – such as the disturbance of hydrological ratios, scouring etc. This fact has been confirmed by the sinkholes which formed to that moment. The lack of the insulating layer above the weathered sandstones caused the infiltration of surface water into the rock mass and, as a consequence, led to changes in the hydrological ratio, resulting in the displacement of the rock material. The roof layers of these voids became fractured and caused collapses travelling to the surface, causing the formation of sinkholes. Because in practice it is impossible to eliminate the infiltration of water, and the goafs resulting from the old shallow exploitation of the 301 seam occur under the entire residential area, it may be concluded that a comprehensive protection of the M. Rej. residential area is not possible even under the condition of precise location of the voids in the rock mass. This is because there are no effective methods of backfilling flooded voids.

Based on the analysis of data obtained in old boreholes and the results of the conducted study, it may be concluded that a large number of voids exists at various depths over the roof of the 301 seam [4]. This implies that it is highly probable that not only primary voids occur in the goafs of the shallow workings, but also secondary voids exist at various depths. In favorable conditions, these voids may contribute to the formation of sinkholes in the future.

The factors that may activate the voids include:

- the effects of mining exploitation,
- changes in hydrological ratios,
- vibrations caused by traffic,
- construction of heavy buildings at the surface.

The presented example of hazard posed by discontinuous deformations indicates that this hazard may not be fundamentally liquidated due to the following reasons:

- there is no practical possibility of detecting all the voids that are present beneath the residential area,
- there is no effective method of backfilling flooded voids,

- there is no possibility of preventing the infiltration of surface water into the rock mass.

Despite the long time that has passed since the shallow exploitation has ended, it may thus be concluded that the hazard caused to the surface by the discontinuous deformations persists and the time of its liquidation may not be determined. In case of conducting future mining exploitation in that area, also the drainage of the rock mass would occur, which would be disadvantageous in view of the hazard posed by discontinuous deformations resulting from the reactivation and displacement of the voids "suspended" in the rock mass towards the surface.

4. Conclusion

In view of the provided examples from two selected hard coal deposits, the following conclusions may be formulated regarding the environmental conditions related to the future development of the deposits:

- the most significant environmental conditions for the redevelopment of a hard coal deposit include the changes of the hydrogeological conditions which have occurred after the mining exploitation has ended, the changes in methane conditions and the reach of the former shallow mining exploitation in view of the hazards posed to the surface by discontinuous deformations,
- the exploitation of coal should only be allowable in the part of the deposit where no former shallow goafs are present or where studies and preventative and protective works were conducted to eliminate the reactivation of the old shallow goafs,
- while examining the redevelopment of deposits within the area of liquidated mines, the actual resistance of surface structures (buildings, lines) to static and dynamic effects of mining exploitation should be considered,
- while determining the resistance category of newly developed structures at areas where liquidated mines once operated, the potential redevelopment of deposits should be considered.

References

[1] Szuflicki M, Malon A and Tymiński M 2016 *Bilans zasobów złóż kopalin w Polsce wg stanu na 31 XII 2016 r.* (Warszawa: Państwowy Instytut Geologiczny)

[2] Chudek M 2010 *Mechanika górotworu z podstawami zarządzania ochroną środowiska w obszarach górniczych i pogórniczych (Gliwice:* Wydawnictwo Politechniki Śląskiej)

[3] Kleta H and Plewa F 2001 Zagrożenie powierzchni terenu po zakończeniu wieloletniej eksploatacji górniczej na przykładzie kopalni „Siersza" *Zeszyty Naukowe Politechniki Śląskiej* seria *Górnictwo* **250**

[4] Kleta H and Zych J 2015 Opinia naukowa dot. Raportu o oddziaływaniu na środowisko przedsięwzięcia polegającego na wydobywaniu węgla kamiennego wraz z kopaliną towarzyszącą ze złoża „Brzezinka 3" Politechnika Śląska Gliwice

[5] Collaborative work 2012 Dokumentacja hydrogeologiczna określająca warunki hydrogeologiczne w związku z zamieszczonym wykonywaniem odwodnienia w celu wydobywania metanu z węgla kamiennego ze złoża „Kaczyce I" według stanu na listopad 2011 roku Państwowy Instytut Geologiczny Państwowy Instytut Badawczy Oddział Górnośląski im. St. Doktorowicza Hrebnickiego Sosnowiec

[6] Collaborative work 2015 Projekt koncepcyjny dotyczący sposobu udostępnienia złoża węgla kamiennego „Morcinek 1" wraz z metanem jako kopaliną towarzyszącą Kraków

[7] Collaborative work 2013 Opinia hydrogeologiczna określająca warunki prowadzenia robót górniczych przez NWR „KARBONIA" S.A. w obrębie obszaru „Morcinek 1" Główny Instytut Górnictwa Zakład Geologii i Geofizyki Katowice

[8] Collaborative work 2015 Raport o oddziaływaniu na środowisko przedsięwzięcia polegającego na wydobywaniu węgla kamiennego wraz z kopaliną towarzyszącą ze złoża „Brzezinka 3" oraz na budowie, prowadzeniu i likwidacji zakładu górniczego „Brzezinka 3" Dokumentacja pracy badawczo – rozwojowej Główny Instytut Górnictwa Katowice

IOP Conf. Series: Earth and Environmental Science **174** (2018) 012005 doi:10.1088/1755-1315/174/1/012005

Application of petrophysical shale gas model for CO_2 storage capacity assessment of coals

Marcin Lutynski[1] and Lucio Zavanella[2]

[1] Faculty of Mining and Geology, Silesian University of Technology, Akademicka 2, 44-100 Gliwice, Poland

[2] Department of Mechanical and Industrial Engineering, Brescia University, via Branze 38, 25123 Brescia, Italy

marcin.lutynski@polsl.pl

Abstract. Carbon dioxide storage in unmineable coal seams is a potential solution for the reduction of GHG emissions. For the purpose of storage capacity estimations various static and dynamic models are applied and in this study a simple petrophysical model, originally designed for shale gas reservoirs, has been used. This model comprises of volumetric (free gas) and surface (adsorbed gas) component. Important input data for the model are Langmuir constants and other reservoirs parameters such as water saturation, porosity and rock density. Sorption isotherms on two distinct coals from the Upper Silesia Coal Basin were measured and calculated Langmuir constants with typical reservoir parameters were implemented into the model. It occurred that coal with higher ash content and higher sorption capacity for methane had a lower carbon dioxide sorption capacity. This was reflected in the overall carbon storage capacity which for the two coals varied from 31 m^3/t to approximately 42 m^3/t of coal at the storage pressure of 5.5-6 MPa where the majority of stored gas is the surface component (sorbed gas).

1. Introduction

Regulations on the GHG emissions in European Union are forcing countries where energy mix is based on fossil fuels to implement actions that would either allow for switching into other energy resources or reduce GHG (mainly CO_2) by other means. It occurs that even geothermal energy production contributes to CO_2 emissions reaching 330gCO_2/kWh in case of Italy as opposed to 940gCO_2/kWh emitted from coal fired power plants [1]. One of such solutions that could work for both the cases (emissions from fossil power plants and geothermal) is the Carbon Capture and Storage (CCS). In this technology CO_2 is captured in stationary emission source, transported and permanently stored underground in geological traps. This technology is well known and although not widely implemented to date, it has a huge potential and when taking into consideration the dominant role that fossil fuels continue to play in primary energy consumption, the deployment of CCS technologies is becoming increasingly urgent [2]. Concerning Europe, a considerable amount of studies has been conducted about the implementation of CCS in unmineable coal seams, with particular attention to Poland (particularly in the Upper Silesian Basin) where there are still considerable amounts of Coalbed Methane reservoirs. In order to assess CO_2 storage capacity the following parameters have to be taken into consideration:

- Sorption capacity of coal (Langmuir parameters)

- Water saturation of coal
- Coal porosity

In order to assess the dynamic properties of CBM reservoir, other parameters, such as permeability, are needed. Typical static models for the assessment of storage capacity take into consideration absolute sorption capacity and the volume of gas that could be stored in fractures and cleats of coals. In case of coals the vast amount of gas is adsorbed in coal matrix whereas only a minor fraction is the free gas in fractures and macropores.

In this study, we apply a petrophysical static model in order to assess the total storage capacity of coalbed. The model proposed by Ambrose [3] considers not only adsorbed gas measured in adsorption experiment and void volume space of fractures and macropores (porosity) but also subtracts from this sum the volume occupied by the adsorbed phase of gas.

2. Materials and methods

2.1. Materials

In the study two coal samples from the Upper Silesia Coal Basin (USCB) in Poland were used. They represented two types of steam coal i.e. good quality coal with low ash content (Coal A) and low quality coal with high ash content (Coal B). Prior to the start of the experiments the coals were dried and the data refer to the dry basis. Since the coal was acquired from active mine it was difficult to obtain data on water saturation. It was assumed that the coalbed will have two different values of water saturation i.e. 15% and 30% (see table 3), which represent typical values for the USCB. Technical analysis of coals used in the study is shown in table 1.

Table 1. Properties of coal used in the study.

Parameter	Coal A	Coal B
Ash content, %	4.66	11.33
Moisture content (as received), %	0.2	2.1
Moisture content (air dried), %	1.38	7.46
Higher heating value, KJ/kg	33376	22060

2.2. Methods

Typical method for assessing gas storage capacity is to conduct sorption experiment. Hence, a sorption isotherm can be established and, based on that, it is possible to calculate Langmuir parameters. For the purpose of the study a manometric sorption setup was used. Detailed description of experimental procedure can be found in [4]. Experiments were conducted in the following order:

1. Helium expansion – determination of void volume of the setup,
2. Gas evacuation – vacuum for 24 hours,
3. Methane adsorption experiment,
4. Gas evacuation – vacuum for 24 hours,
5. Carbon dioxide adsorption experiment.

Temperature of the experiments was kept constant at 35°C which was similar to in-situ conditions and the maximum pressure was approximately 15 MPa for methane experiment and approximately 10 MPa for carbon dioxide. In case of carbon dioxide experiment, the excess sorption values became negative near the supercritical point which is often observed for microporous adsorbent such as coal.

The numerical model chosen for the assessment of gas-in-place for carbon dioxide is a volumetric method proposed by Ambrose [3]. This model calculates the total gas in-place Gst (i.e. the total storage capacity of coal) in terms of:

- a volumetric component, Gf, which involves gas stored in the pore spaces of the coal matrix as free gas; and
- a surface component, Ga, which accounts for the gas physically adsorbed on large surface areas.

IOP Conf. Series: Earth and Environmental Science **174** (2018) 012005 doi:10.1088/1755-1315/174/1/012005

The total gas in-place can be calculated as:

$$Gst = Gf + Ga \tag{1}$$

All three parameters are expressed in terms of sm³ of CO_2 per tonne of coal. Ga is quantified based on the sorption isotherm measurements and Langmuir isotherm model fitting, while Gf combines those results with physical parameters of the reservoir.

$$Ga = V_L \frac{p}{p+p_L} \tag{2}$$

Where V_L is the Langmuir volume (sm³/Mg), p is the equilibrium pressure (MPa) and p_L is the Langmuir pressure (MPa). The free gas component has the following formula:

$$Gf = \frac{32.0368}{Bg}\left[\frac{\varphi(1-Sw)}{\rho b} - \frac{1.318\times10^{-6}M}{\rho s}\left(V_L\frac{p}{p+pL}\right)\right] \tag{3}$$

Where B_g is the gas formation volume factor (-), φ is the porosity of the rock (%), Sw is the water saturation (-), ρb is the bulk rock density (g/cm³); M is the molar mass and ρs is the sorbed phase density (g/cm³).

3. Results and discussion

In figure 1 results of excess sorption measurements (with Langmuir models fitted) of methane are shown. Figure 2 shows results of carbon dioxide excess adsorption and fitted Langmuir model. In table 2 a summary of the Langmuir parameters calculated for the coal samples is given.

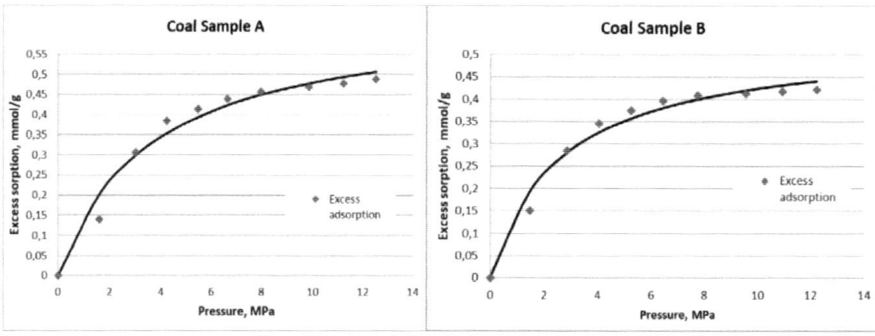

Figure 1. Results of methane adsorption on coal A and B. Black line represents Langmuir model fitting.

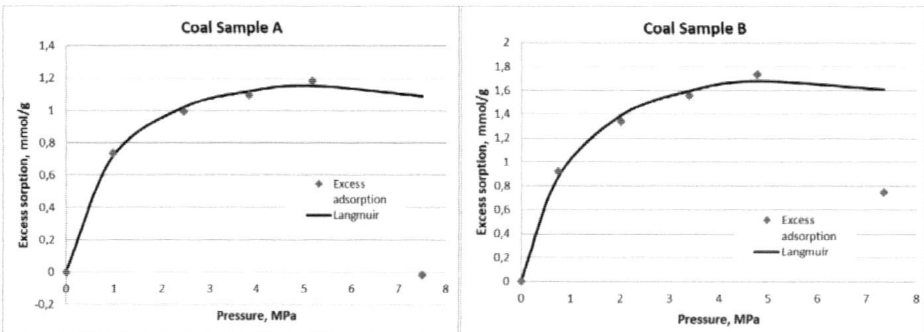

Figure 2. Results of carbon dioxide adsorption on coal A and B. Black line represents Langmuir model fitting.

Table 2. Langmuir parameters calculated for the coal samples.

	Coal sample	Langmuir Volume (sm^3/t)	Langmuir Pressure (MPa)
CH_4	A	14.56	3.57
	B	11.95	2.61
CO_2	A	33.46	1.04
	B	50.24	1.13

Sorption isotherms show that in case of methane sorption capacity is higher for sample A, which can be explained by lower ash and moisture content. For carbon dioxide this is not evident and the maximum sorption capacity of sample B is higher. It can be explained by a slightly higher moisture content since the solubility of CO_2 in water is much higher than of methane. Observation of sorption isotherms is proved by the calculated Langmuir parameters.

The total gas in-place (i.e. the maximum quantity of gas that can be stored in the coal) or CO_2 storage capacity is strictly related to the pressure and temperature of the gas at the moment of injection: for this matter, different injection pressures characteristics of CO_2 storage in coalbed were considered, but the temperature was assumed constant and equal to the one at which the sorption experiments were performed. For the calculation of the total gas-in-place the following parameters, shown in table 3 were used.

Table 3. Parameters used for the calculation of CO_2 storage capacity in coals.

Parameter	Coal A	Coal B
Porosity (ϕ), %	10	10
Water saturation (S_w), %	15	30
ρb g/cm^3	1.38	1.52
ρs, g/cm^3	0.706	
$\dot{M}$, g/mol	16.04	
V_L, sm^3/t	33.46	50.24
p_L, MPa	3.57	2.61

In table 4 results of the total CO_2 storage capacity for two coals used in the study are shown. The assessment was done for two ranges of pressures. Both ranges are considerably low in terms of recommended storage conditions (above 700 m) but could be considered valid for shallow unmineable coalbeds.

Table 4. CO_2 storage capacity results for selected coals.

Storage pressure	5.5-6 MPa		3.5-4 MPa	
	Coal A	Coal B	Coal A	Coal B
Gf, sm^3/t	2,49	0,11	1,44	0,2
Ga, sm^3/t	28,49	42,24	26,52	39,12
% Gf	8,0%	0,3%	5,1%	0,5%
% Ga	92,0%	99,7%	94,9%	99,5%
Gst, sm^3/t	30,97	42,34	27,96	39,32

As could be expected, in both coals a higher amount of CO_2 can be stored with a higher pressure of injection/storage. Coal B accounts for a higher storage capacity at each pressure level compared to coal A, due to the lower adsorption of carbon dioxide in this last type of coal: the difference is considerably high, with the coal B which can store on average 10 sm³/t more than coal A per each pressure range.

Another interesting insight of the results is the breakdown of the total gas in-place value into the volumetric (Gf) and surface components (Ga). The ratio of the first one, which is related to the free gas stored in the pores of the coal, is higher under the same pressure conditions in coal A than in coal B, due to the lower water saturation of the first type of coal. It can also be noticed that the volumetric component tends to increase in coal A with increasing of the pressure, while in coal B it tends to decrease: the reason of these different behaviours is that coal B, adsorbing more CO_2, has a higher increase of the surface component (which is related to the gas physically adsorbed in the coal) with the increase of the pressure than coal A. This proves that in coals a significant part of gas is stored in adsorbed phase rather than in macropores, cleats or fractures. Even considerably higher water saturation of Coal B did not contribute significantly to the lowering of storage capacity. In figure 3 a graphical comparison of results shown in Table 4 is presented. It is evident that volumetric component (Gf) does not have a significant contribution to the overall storage capacity.

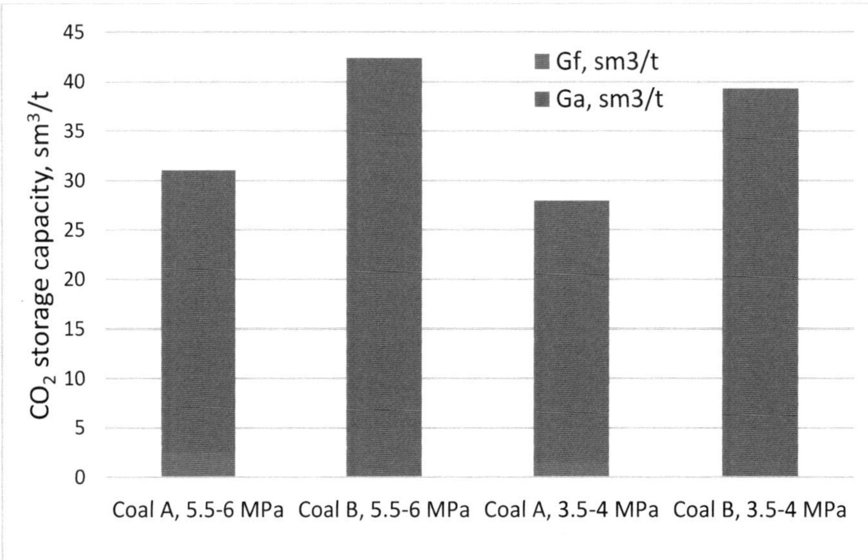

Figure 3. Graphical comparison of calculated CO_2 storage capacity of coals.

In general, the total potential storage capacity can be considered as rather high, reaching over 40 sm^3/t of coal at the pressure of 6 MPa. This high amount results from the method of sorption measurement, which considers ideal conditions and full coal saturation with gas. In practice it is unlikely to achieve such conditions hence obtained data should be treated with caution. The results calculated with the Ambrose model can be used to assess the storage capacity of the reservoir in terms of total sm³ of CO_2 that can be stored. This can be recalculated by multiplying the storage capacity from Ambrose by the density and volume of the coal seam [5]. As density value, the bulk rock density can be considered, while the volume of the reservoir can be assessed by multiplying the area of its surface by its thickness. The area and thickness of the reservoirs were assumed in accordance with typical values for coal seams and assumed equal for both type of coal. In table 5 reservoir parameters and storage capacity expressed in terms of sm³ and tonne of CO_2 are shown.

Results show that a maximum of about 2.5 Gsm³ of CO_2 can be stored in reservoir of coal B at 5.5-6 MPa, while a maximum of about 1.7 Gsm³ of CO_2 can be stored at the same pressure in a coalbed composed by the coal A.

Table 5. Reservoir parameters and storage capacity expressed in terms of sm^3 and tonne of CO_2.

	5.5-6 MPa		3.5-4 MPa	
	Coal A	Coal B	Coal A	Coal B
Area, m^2	4000000			
Thickness, m	10			
ρb t/m^3	1.38	1.52	1.38	1.52
Gst, sm^3	1 709 782 733	2 574 517 421	1 543 128 914	2 390 574 710
ρs t/m^3	1.936			
Gst, t	3 310 891 676	4 985 398 515	2 988 176 554	4 629 204 491

Results show that a maximum of about 2.5 Gsm^3 of CO_2 can be stored in reservoir of coal B at 5.5-6 MPa, while a maximum of about 1.7 Gsm^3 of CO_2 can be stored at the same pressure in a coalbed composed by the coal A.

By multiplying those results by the adsorbed phase density, it is possible to convert the results in terms of tonne of CO_2 stored. Hence, in coal A reservoir a maximum of 3.3 Gt of CO_2 can be stored at 5.5-6 MPa, while in a coal B reservoir under the same pressure conditions approximately 4.9 Gt of CO_2 can be stored.

4. Conclusions

The intention of this study was to perform a simple carbon dioxide storage capacity assessment based on the petrophyscial model designed primarily for shale gas reservoirs. The following conclusions can be drawn based on the obtained results:

- Sorption isotherms show that, as expected, carbon dioxide sorption capacity is higher than for methane; however for the Coal B of lower quality the maximum CO_2 sorption capacity was higher than for coal A,
- In general, the calculated CO_2 storage potential is relatively high and it seems that the model is overestimating the storage potential,
- In order to obtain more credible results it is necessary to conduct additional analysis of coalbed reservoir parameters such as water saturation and, above all, the porosity.

Acknowledgments
Authors of this study would like to thank Sebastiano Corradini for his great contribution to this work.

References

[1] Fridriksson T, Merino A M, Orucu A Y and Audinet P 2017 Greenhouse gas emissions from geothermal power production *Proceedings 42nd Workshop on Geothermal Reservoir Engineering (Stanford California: Stanford University)*
[2] Brickett L, Bromhal G, Miller D and Rodosta T 2016 An update on CCS research and development *Power Eng.* **120**
[3] Ambrose R J, Hartman R C, Diaz Campos M, Akkutlu I Y, Sondergeld C and others 2010 New pore-scale considerations for shale gas in place calculations *SPE Unconventional Gas Conference* Society of Petroleum Engineers
[4] Lutynski M and González G 2016 Characteristics of carbon dioxide sorption in coal and gas shale - The effect of particle size *J. Nat. Gas Sci. Eng.* **28** pp 558–65
[5] Dallegge T A and Barker C E 2000 Coal-bed methane gas in place resource estimates using sorption isotherms and burial history reconstruction: An example from the Ferron Sandstone member of the Mancos Shale, Utah *US Geol. Surv. Prof. Pap.* L1–L26

IOP Conf. Series: Earth and Environmental Science **174** (2018) 012006 doi:10.1088/1755-1315/174/1/012006

Changing the mining industry in the heart of Silesia

Zygmunt Łukaszczyk[1] and Marcin Popczyk[2]

[1] Faculty of Organization and Management, Silesian University of Technology, Roosevelta 26-28, 41-800 Zabrze, Poland
[2] Faculty of Mining and Geology, Silesian University of Technology, Akademicka 2, 44-100 Gliwice, Poland

zygmunt.lukaszczyk@polsl.pl

Abstract: The development of coal mining in the Upper Silesian region began in the seventeenth century. Along with the development of this branch of industry, they began to form the new villages, towns and entire agglomerations in the surroundings. Even before the First World War the Polish mining industry employed over 150 thousand miners. This sector is characterized by a high degree of mechanization of work and technical progress. Miner profession from its inception was set on a pedestal by giving it a number of privileges. Mining industry introduced earliest pension and health insurance, and normalized length of the working day. Expansion of class consciousness, trade unions and political parties, resulted that the Silesia from the start was ahead of other mining and industrial regions at least by one generation. There have passed many generations from bloom years of mining to its current form of activity. At present, the global reorganization and restructuring of the industry hit also Polish mining industry.

1. Introduction

It wasn't that mines were established in centres of cities and villages, but the other way round: cities and villages were built around mines. The development of exploitation of galamine and blende at the turn of the 17th and 18th centuries in Upper Silesia resulted in the necessity to build steel plants. These in turn had to be supplied with coal, which caused the businessmen of the day to commence exploitation of coal in the areas of newly established steel plants so as to avoid unnecessary transportation. In Upper Silesia, new villages, cities and whole agglomerations developed along with the robustly developing heavy industry and the construction of mines. Over 150 thousand miners were employed in the Polish mining industry prior to the First World War [1]. This sector was characterized by a high level of work mechanisation and technical advancement. From its very beginning, the miner's profession was cherished, which resulted in a series of privileges. It was the mining industry that was the first to introduce health and social insurance and standardise the miners' working hours. The expansion of class awareness, labour unions and political parties gave Upper Silesia a head start over other regions by at least one generation since the very beginning. All this was related to ensuring tens of thousands of households a modern housing, trade, transportation, sport and cultural infrastructure. With years, new settlements were established, villages transformed into cities and the area of Upper Silesia was developing into an agglomeration.

These processes are best exemplified by the existence of the Nikiszowiec quarter region along with the Gieshe residential complex around a well-functioning mine.

Jerzy Giza (his German name was Georg Giesche) was the first one to conduct large-scale extraction of calamine and blende. As early as at the turn of the 17th and 18th, together with Kacper Pielgrzymowski (von Pelchrzim), he extracted calamine and transported it to Prussian cities via the Vistula and Odra rivers.

Since the mid-18th century, the "Gieshe" company was run by managers in the name of the heirs, and later, as an enterprise with a legal personality, it transformed into a multi-unit, multi-branch conglomerate encompassing ore mines, coal mines, non-ferrous metals steelworks, as well as factories of porcelain and chemical products such as sulphuric acid and artificial fertilisers.

Thus, it may be said that it has all started with zinc. It should be said that in the past, the Silesia and Dąbrowa Basin was the land of zinc, coal and iron [2]. These products have determined the directions of the dynamic development of the region for hundreds of years.

In the industrial period, in the early 20th century, namely in 1905, the general director of the Giesche heirs corporation – Anton Uthemann, reached the conclusion that a residential settlement should be built for the German miners. The miners who were brought from central Germany were, however, unable to settle and feel at home and they quickly returned home. The Giszowiec colony (initially called Gieschewald) was established in 1907 and was in construction until 1910. Giszowiec was designed by two German designers – Georg and Emil Zillmann (figure 1).

Figure 1. Georg and Emil Zillmann.

The settlement was a "garden city" and the pioneer in such type of development was a British town planner – Ebenezer Howard [3]. The settlement was founded in the middle of the woods belonging to the Janów commune (figure 2).

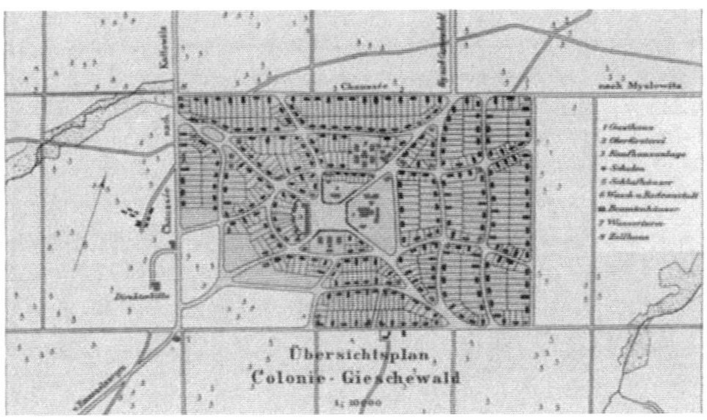

Figure 2. The primary construction plan of the Giszowiec colony.

IOP Conf. Series: Earth and Environmental Science **174** (2018) 012006 doi:10.1088/1755-1315/174/1/012006

Similarly to Giszowiec, the neighbouring Nikiszowiec (figure 3) was established during the rapid industrial development of Upper Silesia in the years 1908 – 1919 and 1920 – 1924. Also this quarter was designed by Georg and Emil Zillman.

The beginnings of Nikiszowiec are strictly related to the "Giesche" coal mine, where construction of two new shafts – "Carmer" (currently "Pułaski") and "Nickisch" (currently Poniatowski) in the "Reserve" panel was commenced in order to exploit new coal seams. The Nikiszowiec quarter took its name from the last director of the Gieshe plant in Katowice – Kurt Nikisch.

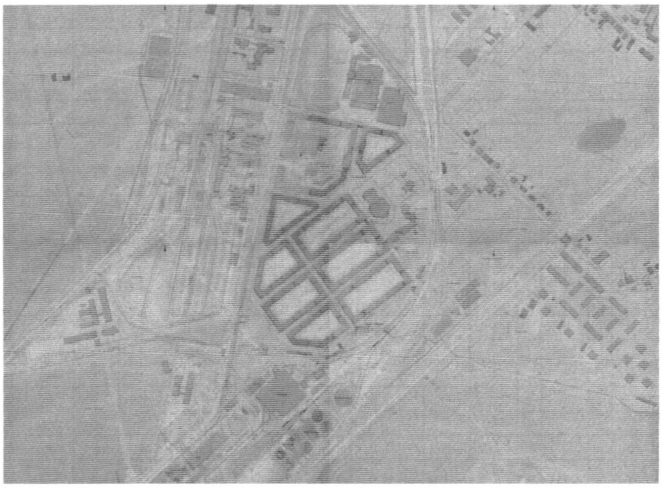

Figure 3. Construction plan of the Nikiszowiec colony.

To commence the exploitation, however, significant manpower was required and bringing so many people was not easy. The area was a woodland where no settlements or villages existed. The development of the residential buildings was very different from the one known from the old Giszowiec.

The plan of the Nikiszowiec settlement encompassed 10 3-floor buildings including 7 residential buildings, and two closed housing and services buildings as well as an open quarter with flats, shops, bathhouse, laundry, workers' hotel and the management of the mine [4].

The Katowice – Murcki (figure 4) quarter was yet another area of Upper Silesia where a miners' settlement was established around a functioning mine. The first written records of extraction of coal in Upper Silesia, from 1657, concern the Murcki coal mine.

Figure 4. Kostuchna 1910.

In the post-war time, the Murcki coal mine was merged with the Boże Dary coal mine and subsequently it continued to operate independently. In the 1970s the mine was once again merged with the Boże Dary mine and the two-section Murcki coal mine was established with the Murcki section and the Boże Dary section. The exploitation of coal in that coal mine continued without interruptions for over 350 years, until 2015 [5].

In the entire post-war period, coal mining was one of the most important sectors of the socialist economy. Simultaneously, it was decisive for the importance of the Silesian region. The incredibly high status of mining in the Polish economy is evidenced by the fact that through nearly the entire period of the Polish People's Republic it had "its own" management at a central level – as early as in 1949, the Ministry of Mining and Power was established, and in 1950, the Ministry of Mining was separated from it and transformed in 1955 into the Ministry of Coal Mining. In the years 1957 – 1976, the Ministry of Mining and Power functioned at the central level, while in the years 1976 – 1981 it was the Ministry of Mining. In the 1980s, the mining industry was managed by the Ministry of Mining and Power. Despite the changes and reorganisation within the 45 years following the war, the coal mining industry had a "visible" management structure at the central level. The central structures, except for a short period in the 1950s, when the plants were subject to the central management of ministries, governed the medium-level management bodies, the so-called "Unions". Their purpose was to organise the individual Upper-Silesian and Dąbrowa Basin's mines.

To exemplify this, the Katowice Union of Coal Industry initially encompassed the following coal mines" „Janów" (later „Wieczorek"), „Kleofas," „Katowice," „Wujek," Mysłowice," and „Eminencja" (later „Gottwald"). In 1968, the Union also encompassed the "Murcki" and "Lenin" (curently "Wesoła") coal mines.

On August 2nd, 1982, by the ordinance No. 164/82, the Unions were transformed into Associations. In Katowice, the Association of Coal Mines encompassed 10 coal mines and 3 auxiliary plants. The mines were a part of the Association for only 2 years. On September 29th, 1984, multi-plant companies were formed, called the Guilds. The Katowice Coal Guild included 10 mines and 3 plants. Another stage in the transformations was the establishment of five multi-plant mining companies at the end of 1988. These companies were called Coal Exploitation Companies [5].

At the onset of the system transformation of 1989, the mining industry was a real power, at least considering the number of persons employed in that sector: 415 740 employees and an output of 177.5 million tons [6].

The political and economic transformation resulting in the switch to a democratic state and free-market economy has had its impact also on this sector. The inefficient structure of the industry (70 mining plants as separate business entities), low capability of generating measures for replacement investments, low work efficiency and the deterioration of the profitability of the coal mining industry has forced the adjustment of the Polish coal mining industry to the conditions of market economy. The processes of these changes were supported – with varying results – by the sector repair programmes implemented in the years 1993-2016, such as [7]:

- March 1993: "The programme for restructuring of the Polish coal mining industry – implementation of the 1st stage within the financial capabilities of the state."
- June 1993: interim adjustment of the restructuring programme: "The programme for the prevention of the bankruptcy of the coal mining industry" – conducted in the period of June – December 1993.
- March 1994: "2nd stage of the implementation of the programme for restructuring of the Polish coal mining industry in the years 1994-1995."
- April 1996: "Coal mining industry – the state's and sector's policy for the years 1996 – 2000. Program for the adjustment of the coal mining industry to the conditions of market economy and international competitiveness." This programme, in a more unambiguous manner, established objectives such as: the profitability of mining, reduction of exploitation costs, and the achievement of an economically rational coal price. Among other methods, these objectives were to be achieved by the reduction of employment by 80 thousand persons and

the decrease in the output by approximately 20 million tons a year. 3 209 million PLN from the budget were allocated for that purpose. It should be added that also this programme has not delivered the assumed effects.

- June 1998: "The reform of the coal mining industry in Poland in the years 1998-2002", Conducted based on the Act dated 26.11.1998 on the adjustment of the coal mining industry to functioning within market economy and the special rights and tasks of mining communes.
- December 1999: adjustment of the government programme. Reform of the coal mining industry in Poland in the years 1998-2002, adopted by the Council of Ministers on 21.12.1999.

The main objectives of this programme included:

- making the mining industry profitable,
- liquidation of the excessive production capacity of the industry – approx. 36 million tons/year,
- reduction in employment by approx. 115 thousand persons by the end of 2002,
- complete liquidation of 15 mines and partial liquidation of 9.

This programme – the most mature one, based on social dialogue – equipped both with financial and legal instruments as well as a rich package of social protection, has brought the expected results.

- April 2004: "Restructuring of the coal mining industry in the years 2004-2006 and the strategy for the years 2007-2010."
- July 2007: "Action plan for the coal mining industry in Poland for the years 2007-2015."
- January 2015: "Repair plan for the Kompania Węglowa S.A."
- February 2015: "Repair programme for KHW S.A." (the last two programmes were not intended for the entire sector).

The conducted activities have brought significant results, which should be viewed in domains such as: technology, production, social aspects, economy and safety. As a result of the many years of restructuring, the following reductions took place [8]:

- number of mining plants – from 70 in 1990 to 22 as of 31.12.2016,
- output – from 147.4 million tons in 1990 to 70.4 million tons in 2016,
- total number of employees – from approx. 388 thousand in 1990 to 84 645 by the end of 2016.

Besides the problems of technical and organisational nature, the restructuring gives raise to human problems – both individual and social. This means the perpetual balancing between the expected economic objectives and the concept of sustainable development. The success of such changes is always dependent on skilful social dialogue.

2. Redevelopment of mining areas

A modern mining plant also means social responsibility and care for the environment, so that the effects of the exploitation would be as little burdensome as possible. The pro-ecological activity encompasses, i.a. the areas of protection of surface water, management of mining wastes, reduction of dust and gas emissions to the atmosphere, reduction of noise emissions as well as the reclamation of mining areas. Many thousands of hectares of these areas were provided to mining communes. These areas served the establishment of many sub-zones of the Katowice Special Economic Zone (e.g. in Żory or Jastrzębie), business activity zones and business incubators (figure 5).

Figure 5. Area of the "Warszowice Panel" Pniówek - area of the cancelled coal mine construction – 39.76 ha.

There are 12 companies in that area. Declared investment: 599 000 000 PLN, actually borne cost: 766 000 000 PLN. Employment: 2105 persons (as of the end of 2016).

The large area of the post-mining grounds and the inspirations of contemporary architects also give many architectural possibilities that preserve a certain aspect of the mining culture.

The joinery (adapted as a museum) and the machine room of the "Głowacki" shaft were made available to tourists, who may now view a unique steam powered hoisting machine from 1900. The historic mine is not only a mining museum. The place is intended for a broadly understood museum and cultural activity. The Ignacy coal mine is the oldest ROW mine. Its establishment dates back to 1792. The nowadays historic mine functioned as KWK Hoym and KWK Hoym-Laura coal mine until 1936. It was closed in 1995 (figure 6, figure 7).

Figure 6. Ignacy mine before revitalisation.

Figure 7. Current view of the Ignacy mine.

To exemplify this – there were areas in the strict centre of Katowice that required redevelopment and that were provided to the city by the Katowicki Holding Węglowy. These were the areas of the former "Gottwald," "Katowice" and "Kleofas" coal mines.

The area of the "Gottwald" coal mine was designated for the construction of the largest shopping centre in Silesia – the Silesia City Center. It was one of the first investments in the region that served the management of a post-mining area. The design provided for the preservation of two of the plant's buildings – they now hold the St Barbara Chapel, the management of the centre, a restaurant, an art gallery as well as the tower of the "Jerzy" shaft, which is the landmark of the centre. Silesia City Center – one of the largest sales, services and entertainment centres in Poland (fifth in view of the area), with a total area of several hundred thousand square meters, out of which 86 000 m^2 serves sales, 100 000 m^2 is a residential area (Dębowe Tarasy estate) and 60 000 m^2 is an office area (Silesia Office Towers) – figure 8.

Figure 8. Silesia City Center.

The area of the former "Katowice" mine that operated in the years 1823-1999 (figure 10) now holds the new seat of the Silesian Museum (figure 9). The historic buildings from the late 19th century are now an integral part of the museum. The centre of Katowice, with a void left by the mine, went through redevelopment in a scale unprecedented in Poland. One of the assumptions was for the city to put emphasis on culture. The facilities established in that zone are supralocal and stimulate the entire region. At the area of that same company, the Polish National Radio Symphony Orchestra has found its seat (figure 11).

Figure 9. Silesian Museum in Katowice.

Figure 10. KWK Katowice mine before redevelopment.

Figure 11. The symphony orchestra's seat.

3. Description of the cultural transformation

The cultural and economic changes proceeded gradually, by having an imprint in the surrounding reality and often drawing new models from it. On the other hand, the hermetic mining culture has provided the changing reality with numerous positive qualities and phenomena. One should mention the high level of social participation in the business processes, the miners' ethos and culture of work, the pragmatism and the professional and social co-resposibility.

For dozens of years, the development of the Silesian agglomeration has contributed to the formation of a region characterised by specific traits to an unprecedented scale in the centre of Europe. Despite the dynamic development of the Silesian agglomeration and the mining industry, the mining culture and traditions are being cultivated and are now an inseparable element of our region. The mining industry, cooperating in full symbiosis with local governments, adjusts to the conditions of the centres of cities concentrated in the heart of Silesia. The most modern investments in Poland were established around the Silesian mines and at their premises [5].

Figure 12. Srebrne Stawy golf course.

The Srebrne Stawy golf course in Bytom is one of the most interesting and the largest redevelopment investments of a post-industrial area in Poland as well as it is the first golf course opened on a post-mining area (figure 12). Also the largest Polish private hospital was built on a post-mining area (figure 13).

Figure 13. The GeoMedical hospital in Katowice.

On the other hand, the development of mines ensured the development of the Silesian agglomeration. The high level of urbanization, the most developed road and railway infrastructure in the country, was the result of the role that was played by the Silesian mining industry for dozens of years. The rule of sustainable development followed by the mining industry consists in supporting the local communities which live by the mining plants since decades and consider the mines their providers. The capital drawn from their tradition is always an investment in the future [9].

References

[1] Mączak A 1981 *Encyclopedia of Poland's Economic History until 1945* ed K. Piesiewicz (Warszawa: Wydawnictwo Wiedza Powszechna)
[2] Kostro T 2004 *In the beginning there was zinc* (Sosnowiec: Wydawnictwo SCW)
[3] Szaraniec L 1984 *Settlements and quarters of Katowice (Katowice:* Wydawnictwo Śląsk)
[4] 2017 *Magical Nikiszowiec, a historic district of Katowice*
[5] https://www.katowice.eu/miasto/turystyka-i-sport/atrakcje-turystyczne/nikiszowiec
[6] Wilczyński M 2013 *The twillight if hard coal in Poland* (Warszawa: Wydawnictwo Fundacja Instytut na rzecz Ekorozwoju)
[7] Turek M 2016 *Key changes in hard coal mining in relation to selected thoughts of the LAUDATO SI papal encyclical* collective work edited by Fr Kania W CARBO ET TERRA LAUDATE DOMINUM (Katowice: Wydawnictwo GIG) pp 107–26
[8] ARP S.A. reports for 2015 and 2016 Katowice
[9] Wuwer A and Szlachta D 2017 *De profundis ad Te clamamus. The world of miners in radio and TV sermons of Katowice's bishops of the St Barbara celebration (1983-2016).* Edition of sources with commentary Theological Department of the University of Silesia Katowice pp 39–45

IOP Conf. Series: Earth and Environmental Science **174** (2018) 012007 doi:10.1088/1755-1315/174/1/012007

The future of hard coal compared to other energy carriers

Anna Manowska and Aurelia Rybak

Faculty of Mining and Geology, Silesian University of Technology, Akademicka 2, 44-100 Gliwice, Poland

anna.manowska@polsl.pl

Abstract. Primary energy sources in Poland are mostly fossil fuels, such as: hard coal, brown coal, natural gas, petroleum and other sources, which include, for instance, renewable energy sources. However, the major role is played by hard and brown coals. For many years hard coal has guaranteed energy security as well as energy independence and the forecast about the demand for net electrical energy [TWh] developed by the Ministry for Energy shows that coal will remain the leading source. Polish mining has been one of the most important industrial sectors for a long time, but for the last several years it has been in some dire straits. One of the main problems of the mining industry is lack of national strategy concerning the future of the fuel and energy sector, especially the energy mix.

1. Introduction

In 1990 the Polish government began the reorganization of hard coal mining in order to improve the financial situation of mines and to increase their competitiveness on the global market. Until the end of 1990 there were 70 operating mines in Poland, which employed 387 000 workers and extracted about 150 million tons of hard coal [1]. Currently, there are 23 operating mines, which employ 80 000 workers and extract about 70 million tons of hard coal [2]. The conducted reforms of coal mining included organizational, employment, financial and property restructuring. The reorganization was implemented in five political stages. The first stage between 1990 and 1997 abolished the centralized management system. In 1993 seven coal companies came into being. Between 1990 and 1997 the further 16 mines were closed down. The second restructuring stage took place between 1998 and 2002. 13 mines were closed down and the employment rate was reduced to 140 000. At the end of 2002 only 41 mines were in operation. The conducted changes in the hard coal mining sector caused an improvement of the financial situation of mines. Between 2003 and 2006 the third stage of hard coal mining restructuring took place. 9 mines were closed down, which left 32 mines with 121 000 employees. The fourth stage of the government's policy concerning hard coal mining restructuring occurred between 2007 and 2011. In July 2007 the government adopted another road map for hard coal mining. The premises of the programme were to reduce the coal extraction costs, to take advantage of cutting-edge mining technology, to manage coal companies effectively and to keep financial liquidity in mines.

2. European energy policy

An additional factor influencing the transformations occurring in the mining sector is adjusting the Polish energy sector to the requirements of the European Union [3]. The new Polish energy policy until 2030 is aimed at creating a common energy policy in the European Union and implementing its main objectives. Figure 1 shows energy policy goals for Europe [4, 5, 6].

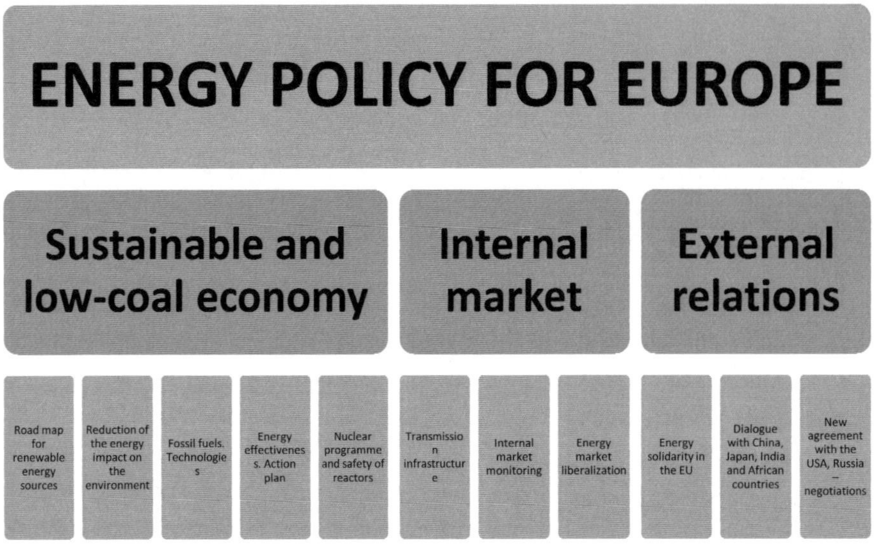

Figure 1. Energy policy for Europe [7].

The basic trends in the national policy are consistent with the main goals of the Community energy policy and include such steps as [7]:

- improving energy effectiveness,
- increasing energy security,
- developing renewable energy sources, including biofuels,
- developing competitive fuels and energy markets,
- reducing the impact on the environment.

The presented trends of the national energy policy fit into the goals of the climate and energy package in Poland, especially through:

- increasing energy effectiveness, treated as a priority which combines the realization of other goals of the new energy policy. The major goals of improving energy effectiveness are:
  - pursuing the zero-energy economic growth, i.e. economic development without a growing demand for primary energy,
  - by 2030 reducing the energy consumption of Polish economy to the EU level 15 from 2005.
- growing use of Renewable Energy Sources (RES). The major goals of the energy policy in this area include [8]:
  - growing use of renewable energy sources in the final energy balance up to 15% in 2020 and 20% in 2030,
  - in 2020 reaching the 10% share of biofuels in the transport fuels market and maintaining the level in the following years,
  - protecting forests from excessive exploitation to obtain biomass and sustainable use of agricultural areas for RES, including biofuels, so as not to cause competitiveness between renewable energy and the agricultural sector.
- the aim of reducing the impact of the energy sector on the environment includes goals and steps to reduce the greenhouse gases emission. The steps to reduce the CO_2 emission are, among others, introducing standards lowering the amount of CO_2 emission per electrical energy unit by 20%, which means that for just this one group the goal corresponds with the premise of the climate and energy package, i.e. the 20% reduction of greenhouse gases emission by 2020 in comparison to 2005, as well as the goals proposed in the ETS Directive of lowering CO_2 emission from the sources included in the ETS System by 21% in 2020 in comparison to 2005 [9].

3. Hard coal in the polish energy balance

The forecast developed by the Ministry for Energy predicts that by 2050 the demand for the energy from hard and brown coals should decrease by 15% and 80% respectively in comparison to 2010 [10]. While analysing the data from 2015, it is clearly visible that hard coal share in energy production has already decreased to the level set for 2050. On the other hand, the brown coal consumption has increased by 12% [11]. Hard coal share in the total demand for energy amounted to 46% in 2015. The predicted values are respectively 43% in 2020 and 33% in 2050. Table 1 presents a forecast developed for energy carriers by 2050. The data are given in TWh.

Table 1. Forecasts of demand for energy carriers by 2050 [12].

	2010	2015	2020	2025	2030	2035	2040	2045
Hard coal	87.9	72.5	76.9	75.9	79	84.4	88.8	82.3
Brown coal	48.6	58.4	53.8	49.6	38.1	11.1	11.3	10.7
Natural gas	6.8	5.8	11.8	11.9	13	18.4	17.5	23.3
RES	11.6	20.6	34	36.9	51.9	61.1	65.1	67.5
Nuclear energy	0	0	0	11.8	23.3	45.1	45.4	44.2
Other	2.6	1.4	1.4	1.4	1.4	1.4	1.4	1.4

Table 1 shows that the demand for hard and brown coals is decreasing, whereas there is a growing tendency for RES, natural gas and nuclear energy. It was assumed that from 2025 the energy mix will also include nuclear energy, whose share will gradually grow. However, these premises should be treated with caution, particularly because until 2017 no decisive efforts were taken in this matter and construction of a nuclear power plant takes about 10 years and costs about 5 million Euros/MW. It is also highly probable that the costs will exceed the predicted level and developed analyses of the investment profitability show that after 2040 nuclear power plants will lose their competitive advantage to wind farms or photovoltaics [13].

Another problem of the fuel and energy sector is the depletion of hard coal deposits, which is shown in table 2.

Table 2. Volume of the hard coal reserves in the active deposits as at 01/01/2015 (mln Mg) [14].

	Balance resources	Industrial resources	Operative resources
The licence duration	14 872	3 230	1 813
For the whole deposit	20 282	6 023	3 390

Taking into account operative resources in the amount of 3 390 million Mg and the current yield level of about 70 million Mg per year, the resource will last for about 48 years, i.e. until 2070, if only we take into account the usual decreasing yield in the final exploitation stage of coal deposits [11]. Obviously, while the coal share in the energy mix decreases, the deposit sufficiency will increase, but at the cost of energy independence. In order to ensure and maintain Poland's energy security, it is necessary to search for alternative solutions providing constant long-term energy supplies. Are renewable energy sources and nuclear energy a solution for Polish energy problems?

4. Renewable energy sources

Renewable energy sources are regulated by the following directives [7]:

- Directive 2009/28/EC of the European Parliament and of the Council of 23 April 2009 on the promotion of the use of energy from renewable sources and amending and subsequently

repealing Directives 2001/77/EC and 2003/30/EC on the promotion of electricity produced from renewable energy sources in the internal electricity market
- Directive 2009/72/EC of the European Parliament and of the Council of 13 July 2009 concerning common rules for the internal market in electricity and repealing Directive 2003/54/EC
- Directive 2005/89/EC of the European Parliament and of the Council of 18 January 2006 concerning measures to safeguard security of electricity supply and infrastructure investment
- The Law on Renewable Energy Sources
- The Directive of the Council of Ministers concerning the sequence of organizing auctions for sale of electrical energy from renewable energy sources in 2017
- The Directive of the Council of Ministers concerning the maximum amount and value of electrical energy from renewable energy sources which may be sold by means of an auction in 2017.

Currently, the renewable energy sources' share is 7% in Poland. The share is constantly growing and is about to reach the level of 15% in 2020. The development of renewable energy sources has an increasing impact on the shape of both the energy policy as well as construction of an electrical energy system [13]. It results from the fact that the EU regulations promote increasingly often the so-called clean energy sources and the climate package is aimed at 'low emissivity' or even 'zero emissivity' [13]. Due to this fact it is necessary to rebuild the Polish National Electrical Energy System, which must be adjusted to receiving energy from dispersed sources.

5. Nuclear energy

The plan to build the first nuclear power plant in Poland is a result of many factors, out of which a growing demand for electrical energy and new environment requirements are the most essential. The Ministry for Energy estimates that by 2030 the demand for electrical energy will have risen by 24%, from the level of 157.7 TWh in 2010 to the level of 206.8 TWh in 2030. Furthermore, the energy and climate package adopted in December 2008, ambitious goals of the climate policy in the European Union and the growing demand for energy resources import along speculative fluctuations of petroleum, natural gas and coal prices make nuclear energy an answer to the previously mentioned issues [14]. It is also necessary to remember about an impact a prospective nuclear power plant may have on Polish energy security, especially considering the uncertainty connected with future natural gas supplies from the east.

The society opposes the energy policy premises adopted by the Polish government, which treats building nuclear power plants as necessary. The first attempt in Poland was undertaken during martial law in 1982, but the society's consistent opposition thwarted the plans. The construction of nuclear power plants in Żarnowiec and Klempicz was suspended. The plans were resumed in 2005. In the next step the Polish Energy Group (PGE) and its daughter company PGE EJ1 planned to build a nuclear power plant between 2009 and 2014. Originally, it was supposed to supply electricity from 2019 [15]. Later the date was postponed several times until it was decided that the power plant was to start operating from 2031. In effect no works on constructing the power plant have begun.

6. Poland's energy structure until 2050

Figure 2 shows the energy structure of Poland until 2050 according to a study carried out by the Ministry for Energy. The study forecasts a gradual decrease in the demand for hard and brown coals. It is predicted that between 2030 and 2050 coal power plants will be gradually closed down. No new mines are being opened. They are replaced by renewable energy sources, nuclear energy and gas. The share of renewable energy sources in the total demand will increase to 33% in 2050. From 2025 nuclear energy will appear in the energy mix. Its share will be gradually increasing until 19% in 2050.

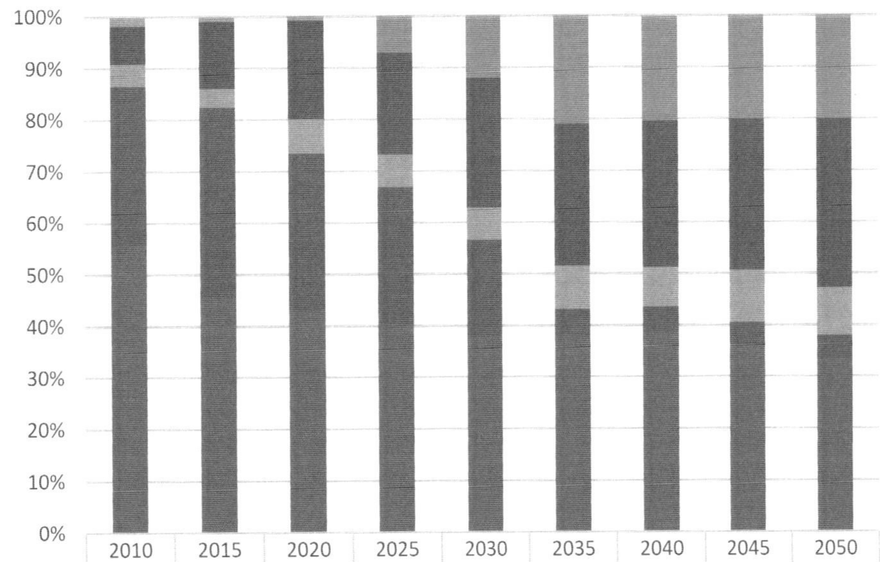

	2010	2015	2020	2025	2030	2035	2040	2045	2050
others	2%	1%	1%	1%	1%	1%	1%	1%	1%
Nuclear binding energy	0%	0%	0%	6%	11%	20%	20%	19%	19%
RES	7%	13%	19%	20%	25%	28%	28%	29%	33%
Gas	4%	4%	7%	6%	6%	8%	8%	10%	9%
Brown coal	31%	37%	30%	26%	18%	5%	5%	5%	5%
Hard coal	56%	46%	43%	40%	38%	38%	39%	36%	33%

Figure 2. Structure of Poland's energy mix [12].

Figure 3 shows a detailed structure of the energy mix highlighting the most important changes occurring in Poland's energy structure.

The energy sector is changing. It is forced by acts and directives of the European Union. The discussed forecasts of the Ministry for Energy undoubtedly show the trends of these changes. Due to a high unpredictability of the energy market, the key issue is an analysis of mega-trends since they determine the direction of the energy development. Not only is it important to adapt to current conditions, but also to plan ahead. Therefore, it is crucial to answer the question whether the objectives will be accomplished within the determined timeframe?

Is it possible to determine if Poland will realize the EU requirements by 2050 only through analysing the potential of national energy sources and the rate of their development?

For this purpose, a statistical analysis has been carried out on historical data concerning the use of individual fuels to produce primary energy and changes resulting from the energy policy have been taken into account. Obviously, it is extremely difficult to capture the national and EU policies mechanisms in modelling, which significantly influences the model's credibility. Therefore, a theoretical model based on linear regression has been built. The model of the development tendency of individual energy carriers is characterized by the ex post error of 8% and Poland's energy structure until 2050 has been based on this model (figure 4).

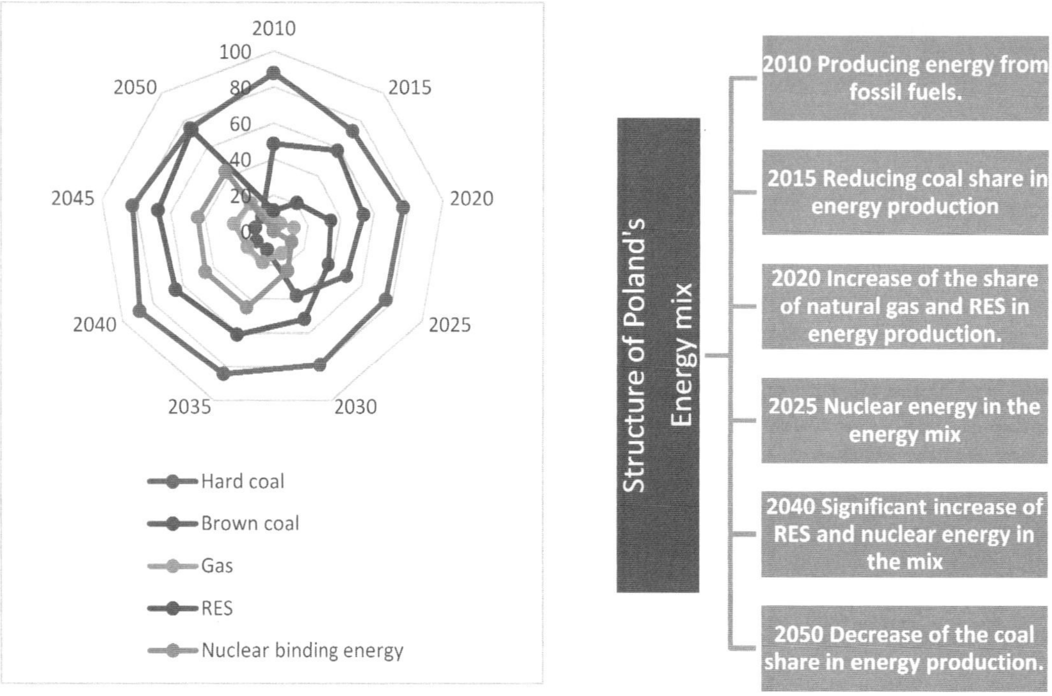

Figure 3. Structure of Poland's energy mix.

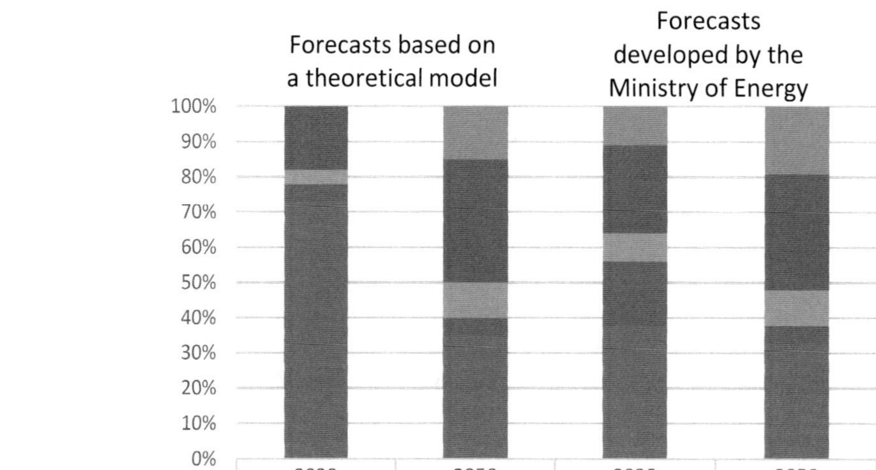

	2030	2050	2030	2050
Nuclear binding energy	0%	15%	11%	19%
RES	18%	35%	25%	33%
Natural gas & oil	4%	10%	8%	10%
brown coal	5%	5%	18%	5%
hard coal	73%	35%	38%	33%

■ hard coal ■ brown coal ■ Natural gas & oil ■ RES ■ Nuclear binding energy

Figure 4. Forecasted structure of energy mix in Poland.

The constructed model undoubtedly proves that the objectives assumed by the Ministry for Energy concerning the energy structure between 2025 and 2050 will not be accomplished. The developed model shows the energy mix compatible with current trends and without undertaking further steps the objectives will not be attained. Problems with achieving the goals connected with nuclear energy and natural gas share are especially evident. The theoretical model shows the natural gas and petroleum shares as imported sources. The import will grow because Poland does not have such resources at its disposal as to be able to meet its own demands. It is also apparent that the energy sector will be mainly based on coal.

7. Conclusions

The Polish energy policy has been in preparation for many years and the proposed directions of actions are changing. Such instability disagrees with the development of the energy sector in Poland. A major part of the discussed steps concentrates on realization and implementation of the European Union policy. The Polish energy policy, apart from following the European Union expectations and implementing the Community regulations, should reach out to global trends in the energy development [7]. Additionally, there are numerous issues on the internal market, which include: running out of national hard coal resources, no effective programmes of developing RES on a larger scale, the necessity to rebuild the Polish National Electrical Energy System or large social reluctance towards nuclear energy. Poland also struggles with the problem of supplies diversification of energy resources, such as natural gas and petroleum, where it is important to create new supplies directions of these resources.

The developed theoretical model shows that the tendency of changes in the energy sector is not coherent with the postulates of the Ministry for Energy. The model solely includes programmes currently in progress but it is a suitable moment to show that it is high time to take necessary steps so as to keep commitments with regard to the energy mix or CO_2 emission. While analysing the forecasts developed for 2030-2050, it is obvious that coal plays a major part in primary energy production. Nevertheless, it is an extremely risky scenario if we consider the system costs, technology costs, problems with covering the national coal resources and increasingly rigorous environment standards.

References

[1] Karbownik A 1997 *Górnictwo węgla kamiennego w Polsce. Ocena stanu aktualnego* ed Bochniarz H and Krajewski S Sektorowe programy restrukturyzacji i prywatyzacja majątku państwowego. Wybór ekspertyz (Warszawa: Zespół Zadaniowy ds. Polityki Strukturalnej w Polsce) pp 80-100

[2] Paszcza H 2017 Górnictwo węgla kamiennego w Polsce w 2016 roku *Wiadomości Górnicze* **6** pp 278–89

[3] Kaliski M, Sikora A P and Szurlej A 2014 Węgiel kamienny w polityce energetycznej Polski *Polityka Energetyczna* **17/3** pp 102-120

[4] Korski J, Tobór – Osadnik K and Wyganowska M 2016 Reasons of problems of the Polish hard coal mining in connection with restructuring changes in the period 1988-2014 *Resources Policy* **48** pp 23-31

[5] Manowska A, Tobór – Osadnik K and Wyganowska M 2017 Economic and social aspects of restructuring Polish coal mining: Focusing on Poland and the EU *Resources Policy* **58** pp 192-200

[6] Jonek-Kowalska I 2018 Method for Assessing the Development of Underground Hard Coal Mines on a Regional Basis: The Concept of Measurement and Research Results *Energies* **11(6)** 1370 DOI:10.3390/en11061370

[7] Bernart R 2011 Polityka energetyczna Polski w aspekcie członkostwa w Unii Europejskiej *Uniwersytet Ekonomiczny we Wrocławiu Wydział Zarządzania, Informatyki i Finansów* **17/3** pp 269-80

[8] Piszczatowska J 2017 *Strategia energetyczna Polski, czyli ile węgla w węglu* access 2017.10.10

[9] Suwała W, Iskrzycki K and Kaszyński P 2011 Dekompozycja redukcji emisji dwutlenku siarki w polskich elektrowniach *Polityka energetyczna* **14/2** pp 107-25

[10] Kamiński J 2014 Zużycie energii pierwotnej w sektorze energetycznym w zależności od struktur rynkowych – podejście modelowe *Gospodarka Surowcami Mineralnymi* **30/4** pp 37-50
[11] Szlązak J and Szlązak A 2017 Dostęp do źródeł energii pierwotnej podstawą bezpieczeństwa energetycznego kraju *Rola polskiego węgla w energetyce krajowej i europejskiej* Kocierz 19 – 21.06.2017 pp 1-21
[12] Ministerstwo Energii 2015 *Projekt Polityki energetycznej Polski do 2050 roku*
[13] Zaleski P 2015 *Perspektywy rozwoju Odnawialnych Źródeł Energii w Polsce*, access 2017.10.10
[14] Kamiński J and Saługa P 2013 Pozyskanie surowców energetycznych na potrzeby wytwarzania energii elektrycznej – koncepcja budowy modelu matematycznego *Gospodarka Surowcami Mineralnymi* **30/1** pp 39-52
[15] Zaborowski T 2012 Czy strategia rozwoju PGE jest niezgodna ze strategią energetyczną Polski? *Energetyka kopalna* access 2017.10.10

Acknowledgments
The work was elaborated in frames of the statutory research [06/030/BK_17/0021]

IOP Conf. Series: Earth and Environmental Science **174** (2018) 012008 doi:10.1088/1755-1315/174/1/012008

Forecast of influences of exploitation conducted in rock mass degraded by the previous activity of the "Bobrek" Coal Mine on the building of historical church in Miechowice

Ryszard Mielimąka and Franciszek Plewa

Faculty of Mining and Geology, Silesian University of Technology in Gliwice, Akademicka 2, 44-100 Gliwice, Poland

ryszard.mielimaka@polsl.pl

Abstract. This paper presents the method of performing forecast of the influence of exploitation planned in seam 503 by the "Bobrek" Unit of "Bobrek-Piekary" Coal Mine on the Holy Cross Church in Bytom - Miechowice. It contains the analysis of church construction and history of its renovations, geological and mining conditions in the church area as well as deformations it underwent as a result of mining exploitation conducted so far. Forecast of influences of planned exploitation was performed with the use of EDBJ program, a formula for subsidences by J. Białek, applying the best local values of parameters of this formula in the sense of the least-squares method. The analyzes presented herein were used to assess the possibility of conducting planned exploitation in the context of protecting this church.

1. Introduction

Forecast of planned exploitation influence on an important public facility of historical character must exhibit considerable credibility due to the need of maintaining the safety of use of such object and preserving its historical values. This is particularly important in the case when this protected object has already been subjected to deformations resulting from long-term, multi-seam mining exploitation conducted in its vicinity. In such case, before performing forecast of influences, it is necessary to put under careful analysis:

- geological conditions in its area,
- previous mining exploitation having an effect on this object,
- construction of the object, its protective measures against the influences of mining exploitation and history of its renovations,
- the size of previous influences of exploitation which it took over on the basis of results of geodetic observations carried out in the object itself and in its vicinity, as well as re-forceasts of influences of previous exploitation.

Forecast of influences of planned exploitation on a protected object is usually performed with the use of geometrical-integral theory formulas utilizing parameters determined locally by fitting subsidences measured in its area to theoretical subsidences by means of the least-squares method. The use of such parameters is of particular importance in the case of rock mass that is already significantly degraded by previous mining exploitation, because at that time one has to deal with a significant random factor in measured subsidence troughs [1, 2]. Forecast of influences performed this way may be the basis for assessing the possibility of taking over forecasted deformations by the protected object.

This article demonstrates the way of performing forecast of planned exploitation influences on a historical protected object. This is presented on the example of the Holy Cross Church in Bytom - Miechowice, which will be subjected to influences of the exploitation of seam 503 conducted by the "Bobrek" Unit of "Bobrek-Piekary" Coal Mine.

2. Description of Church construction

The Holy Cross Church in Miechowice, located in Bytom-Miechowice at 42 Ks. Jana Frenzla Str.,was built in the years 1856-1865 in the neo-Gothic style. It is a brick building made of solid bricks on lime mortar. Finishings and decorative elements are made of sandstone (figure 1).

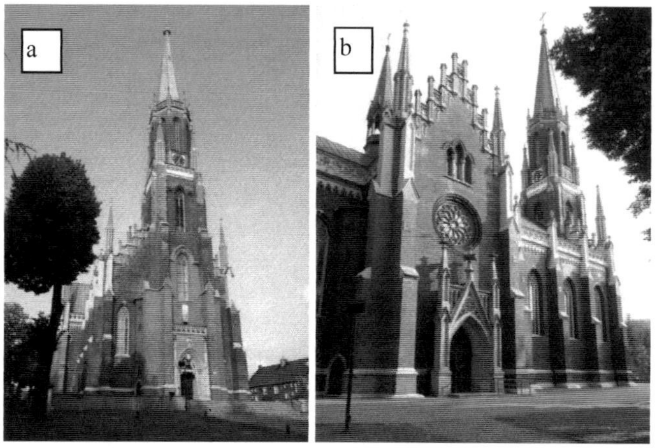

Figure 1. View of the church from the western (a) and eastern side (b).

Plan of the church is organized on a Latin cross-shaped view with the following overall dimensions: church length 51.70 m, width in the transept 30.30 m. In ground floor level the view of the church consists of a main nave with adjoining side naves (northern and southern) in its western part. Main nave crosses with transept, which limits the side naves. In the further part of the church there is a presbytery with a width of main nave. It is adjoined by two sacristies. On the eastern side the presbytery is adjoined by a polygonal apse. The church has a partial basement – under apse and presbytery there is a crypt and boiler room.

Naves of the church are covered with cross-rib vaults. Transept, presbytery and apse – with a star-shaped vault with ribs. Main nave has a span of 8.60 m (in the axes of columns) and side naves are 4.60 m each. Transept is covered with vaults with columns spaced 8.50 m away from each other. In the longitudinal direction vaults are repeated four times counting from the entrance (western side). The axial spacing of supports in this direction is about 5.35 m. Main nave is extended by the presbytery (by about 6.00 m), and then by the apse (by 5.20 m). In the axes of columns there are arches in longitudinal and transverse direction. Vaults placed on arches are at the highest point above the ground + 15.2 m high in main nave, transept and presbytery, and +13.2 m high in side naves. Presbytery has been separated from apse by a gable wall with richly decorated attic. On the transversal extension of arches along the entire height of side walls there are buttresses with width of 0.95 m and length of 1.42 m. In the side walls of the church and apse there are large ogival windows with stone traceries.

The roof of a rafter-and-truss structure is placed on external walls and on the extension of internal columns. The surface of the roof is multi-hipped – gable along the main axis of the church and gable over transept. The roof over apse is multi-hipped. At the point where ridges of main nave and transept cross there is an octagonal spirelet with a height of about 13.2 m. Its spire is at the height of +36.1 m. The roof is covered with copper sheet on full boarding.

Tower of the church is made of brick on the projection of a square. Dimensions of the tower in axes on the ground floor level are about 5.25 m. Projection of the tower partially enters the light of the first span of main nave from the western side. From the level of +28.45 m, the projection of the tower changes shape to octagonal. The brick part of the tower ends at a height of +40.18 m. Above there is a wooden pyramidal roof structure that reaches a height of +53.90 m. It is topped by the spire with a cross. The roof is covered with copper sheet.

As part of major renovations of the church that took place after damage caused by the exploitation of subsequent seams, a number of works have been carried out to increase the church's resistance to the influences of mining exploitation and to ensure the safety of its use. The most important of them include:

a. after damages that occurred until the end of 1953:

- making benches of ferroconcrete under the floor in longitudinal and transverse direction to the church axis with pillars to transfer compressive and tensile forces,
- making ferroconcrete beams in the transept at the level of + 5.00 m to protect the vault of transept from compression,
- making steel tie rods at the level of column capitals in longitudinal and transverse direction,
- making steel tie-rods in external walls at the level of column capitals.

b. in the year 1976 making a horizontal ferroconcrete shield in the shape of an ellipse stiffening the church floor at the ground floor level,

c. the following works after the year 2000:

- laying plastic mesh and applying a gypsum plaster on the upper surface of vaults,
- applying rigid struts between heads of the pillars simultaneously with the already used tie-rods protecting the vaults against the influence of concave curvature and introduction of diagonal tie-rods,
- introduction of ferroconcrete diaphragm to secure the crypt under presbytery and apse against the influence of compressions and against displacements of the column supporting the palm vault.
- dismantling an attic of the wall between presbytery and apse up to the presbytery roof level due to its large deflection.

Inspection of object's damage carried out in April 2017 [3] showed scratches in walls and vaults as well as in places after earlier rebuildings present mainly in the eastern part of the church. The width of most of the scratches did not exceed a few millimeters.

3. Geological and mining conditions

The "Bobrek" Unit of "Bobrek-Piekary" Coal Mine is located within the Bytom Trough. In the range of recognition, rock mass is built of layers of overburden belonging to Quaternary and Triassic as well as Carboniferous forming the deposit. Triassic and Quaternary deposits reach a total thickness of 185 ÷ 260 m.

Quaternary is represented by gravels, sands, loams, slits, mudstones and clays with pebbles and rock blocks, interbedded with sands. In the area of the Holy Cross Church the thickness of Quaternary layers equals to about 10 m.

Triassic formation, deposited in the form of a trough on Carboniferous, is represented by bunter sandstone and shellbearing limestone. From lithological point of view, it is dominated by limestones, dolomites and marls as well as loams with interbeddings of sand and sandstone. In the floor part of ore-bearing dolomites, there is a zinc-lead ore deposit, which was the subject of mining exploitation but not in the immediate vicinity of the church. The thickness of Triassic formations ranges from 60 m in the south to 200 m in the north.

Deposits of productive Carboniferous are represented in this area by Ruda layers (group of seams 400 from seam 404 to seam 419), saddle layers (group of seams 500 from seam 501 to seam 510) and

Poręba layers. These deposits are composed of interbedded layers of claystones, siltstones, sandstones and coal seams. The thickness of Ruda layers is approx. 570 m, and of saddle layers – approx. 240 m.

Carboniferous layers fall beneath the Miechowice District towards the south, to the bottom of the Bytom trough, at an angle of up to 10 degrees.

Mining exploitation of hard coal deposit under the church in Miechowice District encompassed 11 layers (in 9 seams) including 9 layers exploited with a fall of roof. The depth of this exploitation ranged from 270 m (seam 406/4) to 820 m (seam 510), and the height of longwalls ranged from 1.5 m to 2.2 m.

The main way of exploitation was the longwall system with a fall of roof. At the beginning, in upper-Ruda seams (406/4 - 409) dry backfill was used (in the years 1948 - 1954) while from 1965 onwards the prevailing method of exploitation became the one with a fall of roof. Also – to a small extent – hydraulic backfill was used (in seams 418 and 501).

In 2017, the exploitation is carried out in seam 503 with longwalls 5 and 6.

The exploitation schedule influencing deformations of the church building is shown in table 1, which contains data on exploitation time and roof management system.

Figure 2 shows the location of the edges of exploitation fields extracted from the year 1965 in individual seams – in relation to the Holy Cross Church.

Table 1. Schedule of accomplished exploitation in the area of the Holy Cross Church in Miechowice.

Years of exploitation	1948 1951	1951 1953	1966 1973	1976 1978	1978 1980	1983 1986	1989 1992	1993 1995	2000 2002	2003 2004	2013 2015
Seam	406/4	407/2	414/3	418	419	501	507wg	509wg	509wd	510wg	510wd
System of exploitation	Dry backfill	Dry backfill	Fall of roof	Fall of roof	Fall of roof	Fall of roof	Fall of roof	Fall of roof	Fall of roof	Fall of roof	Fall of roof

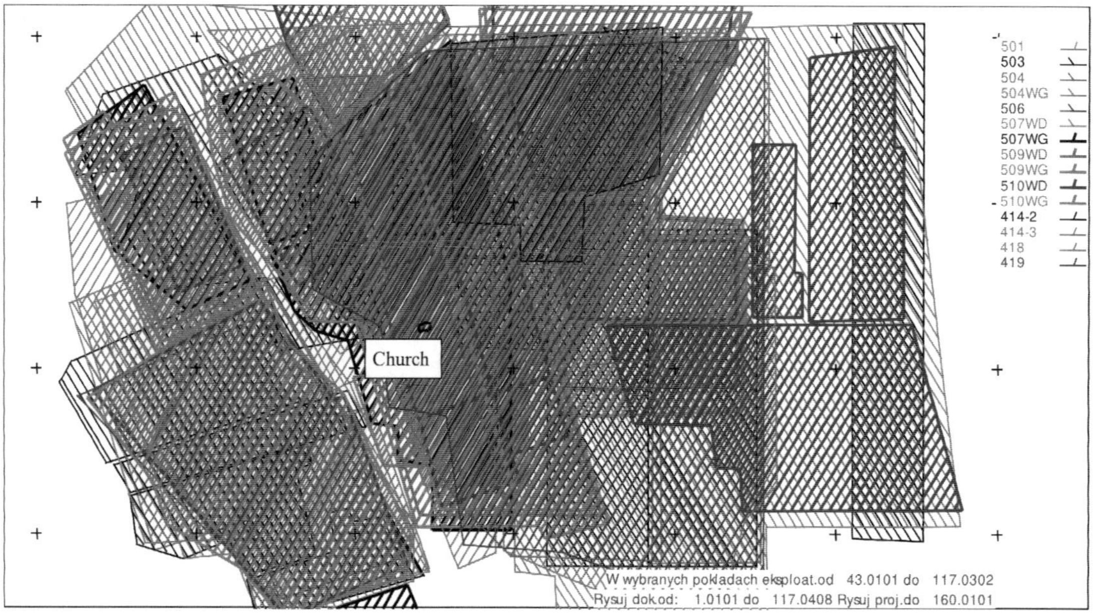

Figure 2. The location of the edges of exploitation fields.

4. The influence of accomplished exploitation on the church building

4.1. The influence of accomplished exploitation in the light of geodetic measurements

Systematic geodetic observations of deformations of the church and its surroundings have been conducted since the year 1965. They include elevation and linear measurements of the measurement network consisting of points located on the walls, pillars and the ferroconcrete band around the church building. Moreover, levelling measurements of the floor are performed inside the church.

The results of geodetic measurements allow to state that the western part of the church building (in the area of portal) has decreased since the year 1927 by 19,40 m (figure 3). A significant increase of subsidences began after the year 1965 in connection with the exploitation of lower- Ruda seams (414/3, 418 and 419).

Summary values of subsidences from 15.01.2014 to 28.02.2017 range from 1628 mm to 2003 mm, while the increase in the difference of subsidences over the length of the church amounts to 68 mm.

The levelling measurement of the church floor made on 15.02.2017 shows that inclinations along the length of the church are generally directed eastward and range from 13.2 mm/m to 35.3 mm/m. In the perpendicular direction they run to the north and amount up to 5.7 mm/m.

Vectors of the church tower deflection, according to the last measurement performed on 28.02.2017, amount to:

- at an altitude of 59.3 m – 17.9 mm/m to the east and 5.2 mm/m to the north (the resultant deflection vector is 18,6 mm/m),
- at an altitude of 28.8 m – 23.0 mm/m to the east and 3.5 mm/m to the north (the resultant deflection vector is 23.3 mm/m).

Subsidences of points in a concrete slab (band) are identical as in the case of points on the church building. The largest subsidence for the period from 14.05.2013 to 03.01.2017 occurred in the eastern part of the band, where it reached almost 2.0 m. In the western part, the subsidences amount to 1.45 – 1.55 m. Correspondingly, the remaining deformation indicators amount to:

- the largest inclinations of ± 12.0 mm/m in the western part of the church,
- the extreme horizontal deformations in time from +1.6 mm/m to -1.2 mm/m in its eastern part.

The analysis of measured deformations indicates that damages to the church building result mainly from the developing of terrain curvatures, because the band made of ferroconcrete limits the influence of horizontal deformations on this object.

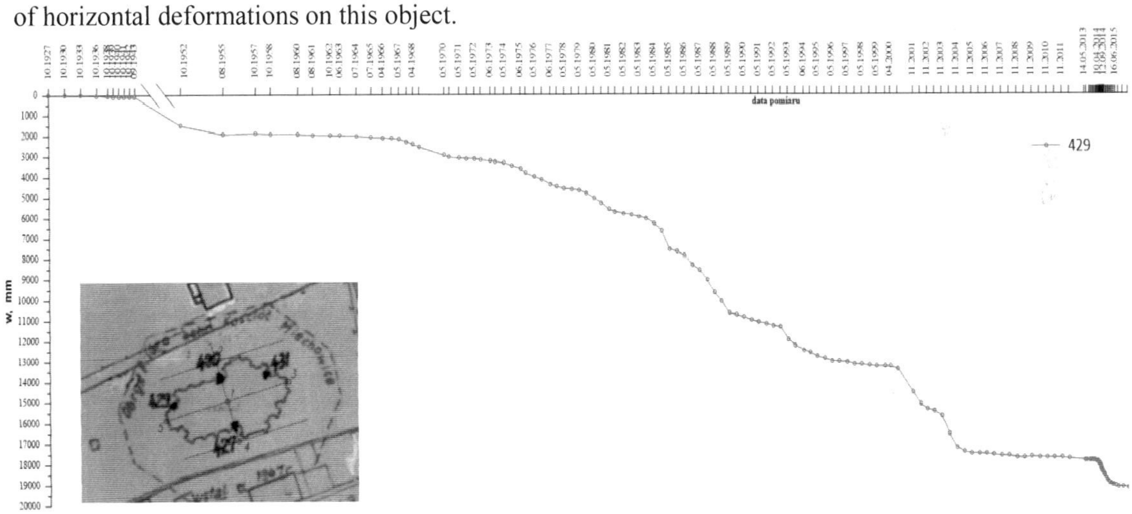

Figure 3. Subsidences of benchmark no. 429 located on the western side of the church from the year 1927 [4].

4.2. The influence of accomplished exploitation in the light of re-forecast of influences

Re-forecast of influences on the Holy Cross Church area was performed for exploitation accomplished since the year 1965 (exploitation in seams: 414/2, 414/3, 418, 419, 501, 503, 504, 506, 507wg, 507wd, 509wg, 509wd, 510wg and 510wd – figure 2).

The analysis of influences shows that exploitation conducted since the year 1965 by the "Bobrek" Unit caused deformations of the church building with the following values:

- increase in subsidences – from approx. 16.90 m on the western side to approx. 17.50 m – on the eastern side,
- increase in inclinations – from approx. 12.5 mm/m on the eastern side to approx. 16.3 mm/m on the western side; these amounts fall within the IV and V inclination categories,
- increase in main deformations – from approx. -6.70 mm/m on the eastern side to about -8.7 mm/m on the western side,
- increase in main curvatures – from approx. +70.8 $m^{-1}10^{-6}$ (radius of curvature 14.1 km) on the eastern side to approx. + 97.2 $m^{-1}10^{-6}$ (radius of curvature 10.3 km) on the western side.

5. Determination of local parameters of the theory of influences

The values of parameters a, tgβ and A_{obr} of the formula for subsidences by J. Białek [5, 6] necessary to evaluate the influences caused by the planned exploitation of the "Bobrek" Coal Mine on the Holy Cross Church in Miechowice, were determined using the EDBJT program, developed in the Department of Deposits Exploitation of the Silesian University of Technology [7, 8]. The values of these parameters are determined by minimizing the residual variance of measured subsidences and the ones calculated theoretically.

Calculations were performed with the use of subsidences of wall benchmarks on the church building which were measured in the period from 15.01.2014 to 28.02.2017.

During the measurement period, twin longwalls with cavings were extracted in seam 510wd: 30a and 31a. Currently, the "Bobrek" Mine conducts exploitation of twin longwalls with cavings, 5 and 6, in seam 503. The location of the edge of this exploitation in relation to the church building is shown in figure 4.

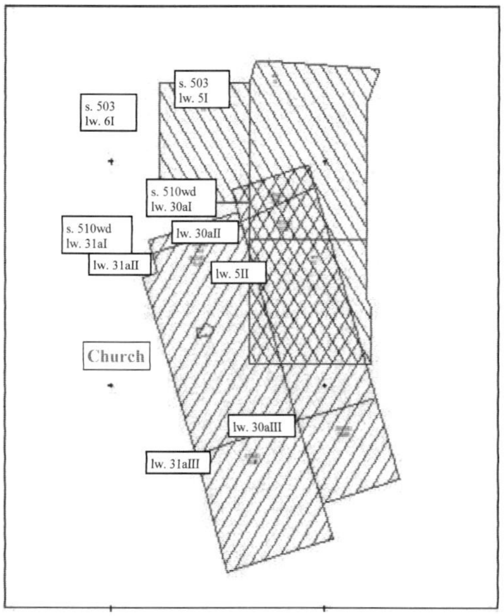

Figure 4. Location of the edges of longwalls extracted in the period from 15.01.2014 to 28.02.2017 in seams 510wd and 503 in relation to the Holy Cross Church building.

Longwalls 30a and 31a in seam 510wd were under exploitation from March 2013 to March 2015 and from June 2013 to July 2015 respectively. These longwalls, exploited in the north-south direction, were set up at an average depth of about 800 m. Their heights were 2.1 m in the panel length sections located in the vicinity of the church building (sections: 30aII with the panel length of about 490 m and

31aII with the panel length of about 460 m) and 2.4 m in the remaining panel length sections (sections: 30aI with the panel length of about 480 m, 30aIII with the panel length of about 200 m, 31aI with the panel length of about 400 m, 31aIII with the panel length of about 280 m).

Exploitation of longwalls 5 and 6, extracted from north to south in seam 503, was commenced on 01.03.2016 and on 01.09.2016 respectively. Until 28.02.2017 longwall 5, located on the eastern side of the church building, reached the panel length of about 670 m and on that day its front was about 40 m south of the church building, while longwall 6 was extracted to the panel length of about 275 m, i.e. its front was situated approximately 265 m north of this building. Longwall 5, to the panel length of about 390 m, was exploited to a height of 2.3 m (section 5I), while on the further panel length the height of exploitation gate was reduced to 2.0 m (section 5II). The height of longwall 6 on the previously extracted section was 2.3 m. The extracted sections of longwalls 5 and 6 are set up at an average depth of approximately 670 m.

As a result of calculations with the EDBJT program, the following values of theory of influences parameters have been determined:

- parameter $tg\beta = 2.80$;
- subsidence coefficient $a = 0.96$;
- peripheral parameter $A_{obrz} = 0.07$.

Matching of theoretical subsidences to the measured ones is characterized by a standard deviation $\sigma = 30.2$ mm and a coefficient of correlation $R = 0.9942$.

The determined values of parameters are local values that can be used only in forecasts of mining exploitation influences on the church building and the surface of terrain in its area.

Values of theory parameters determined on the basis of subsidences of selected points positioned on the church are significantly higher than their values assumed by the "Bobrek" Coal Mine in forecasts of mining exploitation influences on the terrain surface and objects. This phenomenon is present due to the fact that the extraction of discussed longwalls took place in rock mass that was already significantly disintegrated by the previous mining exploitation. In such conditions, the subsidence troughs that emerge on the terrain surface are characterized by a large value of random factor, which means that local inclinations of slopes may be significantly greater than those calculated theoretically with the use of parameter values determined for entire profiles of these troughs.

6. Planned exploitation and its influence on the church building

In the area of the Holy Cross Church in Bytom the "Bobrek" Coal Mine conducts exploitation of longwalls 5 and 6 in seam 503. Longwall 5 was launched in March 2016, longwall 6 in September 2016. Target panel length of longwall 5 is 1309 m, and longwall 6 – 950 m. Depth of exploitation amounts to 670 – 700 m.

Considering the protection of surface during the exploitation of longwalls 5 and 6, it is planned to use mining prevention of limiting the height of longwall 5 to 2.0 m on the 500-meter-long section of panel length (counting from the beginning of longwall on the panel length from 390 to 890 m) and longwall 6 on the section of panel length from 340 m to 950 m. In addition, the speed of front of longwalls 5 and 6 on the panel length from 390 m to the end of longwalls should not exceed 4 m/day (figure 5).

Forecast of exploitation influences of longwalls 5 and 6 in seam 503 was performed with EDBJ program, using formulas of S. Knothe – W. Budryk theory with extensions proposed by J. Białek – the theory parameter values were determined for the Holy Cross Church area.

The analysis of forecast of influences results allows us to state that exploitation of longwalls 5 and 6 will cause the church to suffer deformations with the following values:

- subsidences – from 1374 mm on the western side of the church to 1643 mm on the eastern side of the church,
- final changes in inclinations – from 3.80 mm/m on the eastern side of the church to 6.70 mm/m on the western side of the church; these values fall within the IV and V inclination categories,

IOP Conf. Series: Earth and Environmental Science **174** (2018) 012008 doi:10.1088/1755-1315/174/1/012008

- main tensile deformations extreme in time – from 2.93 mm/m on the southern side of the church to 3.05 mm/m on the western side of the church; these values fall within category II and on the border of categories II and III of horizontal deformations,
- main compressive deformations extreme in time – from -3.54 mm/m on the western side of the church to -4.12 mm/m on the southern side of the church; these values fall within category III of horizontal deformations,
- main negative curvatures extreme in time – from -42.5 $m^{-1}10^{-6}$ (radius of curvature 23.5 km) on the western side of the church to -43.6 $m^{-1}10^{-6}$ (radius of curvature 22.9 km) on the northern side of the church; these values fall within the range of category I of curvatures,
- main positive curvatures extreme in time – from 46.6 $m^{-1}10^{-6}$ (radius of curvature 21.4 km) on the western side of the church to approx. 66.4 $m^{-1}10^{-6}$ (radius of curvature 16.6 km) at the central point of the church; these values fall within the range of category I and II of curvatures.

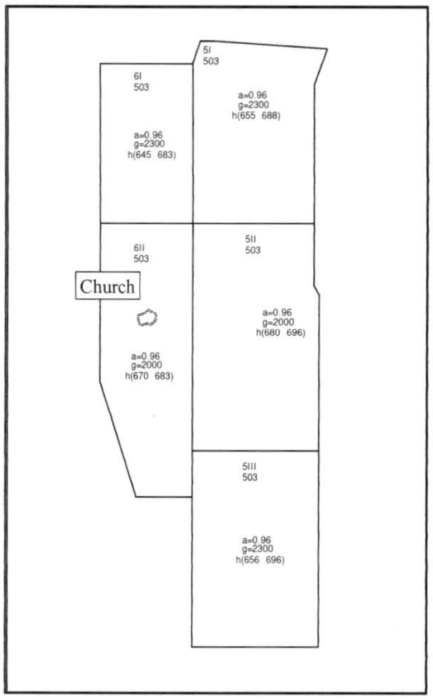

Figure 5. Location of the exploitation edges of longwalls 5 and 6 in seam 503 in relation to the church building

7. Conclusions

Analyses concerning the influences of accomplished and planned exploitation of the "Bobrek" Coal Mine on the historical Holy Cross Church in Bytom – Miechowice, included in this paper, allow formulation of the following general statements:

1. In the past, the church was subjected to the influence of long-term, multi-level mining exploitation, which led to numerous damages, that were successively repaired. During major renovations, the church's construction was strengthened by installing a steel grate in the level of vault support (column heads) and a ferroconcrete shield with ferroconcrete tie-rods below the church floor.

2. The analysis of the results of cyclic geodetic measurements conducted at points placed on the walls and pillars of the church, and also on a ferroconcrete band around the church revealed, that damages occurring in this building are mainly caused by curvatures emerging as a result of mining exploitation in its area. Stiffening the ground floor of the church with a ferroconcrete shield together

with ferroconcrete tie-rods significantly limited the influence of horizontal deformations on this object.

3. Parameters of the theory of influences determined on the basis of matching theoretical subsidences to subsidences measured in the area of the church are significantly larger than those commonly accepted. Because of considerable and diverse disturbance of rock mass the value of parameter a can locally become close to 1.0 and the value of parameter tgβ may get close to 3.0. Therefore, it is not appropriate to perform the forecast of influences for important buildings using standard parameter values because it significantly increases error of the determined deformation values.

4. The stiffening of the church on the ground floor level fulfills its task and is beneficial in terms of assessing the possibility of taking over influences of exploiting seam 503 with walls 5 and 6 situated at a depth of approximately 700 m by this building. With such depth of exploitation extreme curvatures in fact fall within one category higher than extreme horizontal deformations.

5. In terms of protecting the church building and heavily built-up terrain surface in its area it is important to conduct mining prevention of reducing the height of longwalls 5 and 6 in a substantial part of their panel lengths to 2.0 m as well as limiting their daily progress to a maximum of 4.0 m and maintaining its stability. This way of conducting mining exploitation will significantly reduce the size and increments of deformations on the terrain surface.

6. Numerical analysis of exploitation of longwalls 5 and 6 in seam 503 influence on the church construction performed at the Faculty of Civil Engineering of the Silesian University of Technology [9] showed the possibility of emergence of further scratches in the church structure. These scratches will not pose a threat to this structure due to steel reinforcement located in the level of vault supports.

References

[1] Mielimąka R 2009 *Wpływ kolejności i kierunku eksploatacji prowadzonej frontami ścianowymi na deformacje terenu górniczego* (Gliwice: Wydawnictwo Politechniki Śląskiej)

[2] Strzałkowski P 2010 Forecasts of mine-induced land deformations in consideration of the variability of the parameter describing the process kinematics *Archives of Mining Science* **55/4** pp 865-72.

[3] Szojda L, Kubica J and Kotala B 2017 Analiza numeryczna wielokrotnej eksploatacji pokładów węgla na murowaną konstrukcję sklepień kościoła *Konferencja Naukowo-Szkoleniowa „XIV Dni Miernictwa Górniczego i Ochrony Obiektów Budowlanych na Terenach Górniczych"* Ustroń 27-29 września 2017

[4] Kowalski A 2017 Ekspertyza górniczo-budowlana dotycząca wpływu zwiększenia wybiegu ściany nr 6 w pokładzie 503 przez Węglokoks – Kraj KWK Bobrek-Piekary na obiekty zabudowy powierzchni w szczególności kościoła pod wezwaniem Św. Krzyża w Bytomiu Miechowicach GIG Katowice

[5] Białek J 2003 *Algorytmy i programy komputerowe do prognozowania deformacji terenu górniczego* (Gliwice: Wydawnictwo Politechniki Śląskiej)

[6] Białek J 1991 Opis nieustalonej fazy obniżeń terenu górniczego z uwzględnieniem asymetrii wpływów końcowych *Zeszyty Naukowe Pol. Śl. seria Górnictwo* **194**

[7] Białek J and Mielimąka R 1999 Możliwości zwiększenia dokładności prognoz deformacji terenu górniczego wykonanych przy użyciu powszechnie stosowanych programów komputerowych *Zeszyty Naukowe Pol. Śl. seria Górnictwo* **239**

[8] Białek J and Mielimąka R 2001 Próba weryfikacji parametrów teorii prognozowania wpływów eksploatacji na teren górniczy *Zeszyty Naukowe Pol. Śl. seria Górnictwo* **250** pp 69-79

[9] Kotala B, Mielimąka R and Szojda L 2017 Ekspertyza górniczo-budowlana dotycząca wpływu zwiększenia wybiegu ściany 6 w pokładzie 503 na budynek kościoła pw. Świętego Krzyża przy ulicy Frenzla 42 w Bytomiu Politechnika Śląska Gliwice

The use of modern IT tools to acquire information about geographical space from archival maps

Aleksandra Mierzejowska

Faculty of Mining and Geology, Silesian University of Technology, Akademicka 2, 44-100 Gliwice, Poland

aleksandra.mierzejowska@polsl.pl

Abstract. Rapid technological development observed for many years in Geographic Information Systems (GIS) has enabled the application of this system in many areas of life, including: administration, infrastructure management, in environmental protection, spatial planning, in real estate management, in crisis planning and in many branches of industry, including mining. According to [1], the Geographic Information System should be understood as a system for acquiring, processing, verifying, integrating, manipulating, analyzing and presenting data that are spatially referenced to the Earth. It usually includes a spatial database and appropriate software. The article presents an example of obtaining spatial data from archival maps, which are an important, and sometimes even the only source of information enabling tracking changes occurring in the geographical environment and determining the location of objects whose time of using has come to an end. Using the GIS class software, the author made a multivariate calibration of the archival situational and altitude map of the area of the inactive "Saturn" mine. On the basis of the obtained results, she indicated the most appropriate method for this type of task. The map prepared in this way was used to locate liquidated mine excavations connected to the surface. Acquiring reliable information about the location of this type of excavations is particularly important because of the risk associated with the possibility of outflow of gases adversely affecting the atmosphere and thus health of people staying near them (mainly methane and carbon dioxide) as well as due to the possibility of occurrence of so-called discontinuous deformations.

1. Introduction

Among the modern IT tools used more and more often, also in the mining industry [2, 3, 4, 5, 6, 7], GIS class software can be exchanged. This is called any program that allows entering, collecting, analyzing and visualizing data that are spatially referenced to the Earth. The software is, next to people, equipment, data and procedures of their processing and sharing, the basic element of geographic information systems (GIS - Geographic Information Systems). An example of this type of software is the QGIS program. It is an open, free and cross-platform GIS software (it works on GNU / Linux, Unix, Mac OSX and MS Windows platforms). It was created relatively recently (in 2002) and since then it has been dynamically developing [8, 9, 10].

Geographical information systems allow to collect huge amounts of data, both spatial and descriptive and in various forms (raster and vector), verify them, manage them and conduct various types of analysis (http://gisplay.pl/gis/programowanie-gis.html).

Despite the technological development related to, among others, with such fields as photogrammetry, remote sensing and GNSS measurements, very often an important role in obtaining data supplying geographical information systems is still used in traditional maps. Sometimes, they are a source of information without which it would be impossible to perform analyzes related to, for example, the transformation of the environment. In the case of mining maps, they are particularly valuable in a situation when the mining plant is liquidated, and thus objects associated with their activity. As part of the work, the author used an archival mining map covering the area of the liquidated "Saturn" coal mine. Using the QGIS program, there were made several variants of this map calibration, choosing the best option in its assessment. Such calibrated map and information obtained from available services (Google maps) allowed the author to determine the location of mining excavations located in this area and having connection with the surface in the current spatial situation. Knowledge on this subject will increase the level of safety of users of this area because the presence of these excavations is associated with the risk of discontinuous deformations and the possibility of gas outflow (methane and carbon dioxide) [11, 12, 13, 14].

2. Characteristics of the area of the study

The study covered the mining area of the "Saturn" mine. This coal mine was created from the merger of three other mines: the "Czerwona Gwardia" mine, "Milowice" and "Czeladź". The first one was the "Milowice" mine (1822), then the "Czeladź" mine (1860) and the last of these three, the "Saturn" mine, which in 1951-1984 functioned under the name "Czerwona Gwardia". The mining area included the city of Czeladź, part of the city of Sosnowiec and Będzin (figure 1).

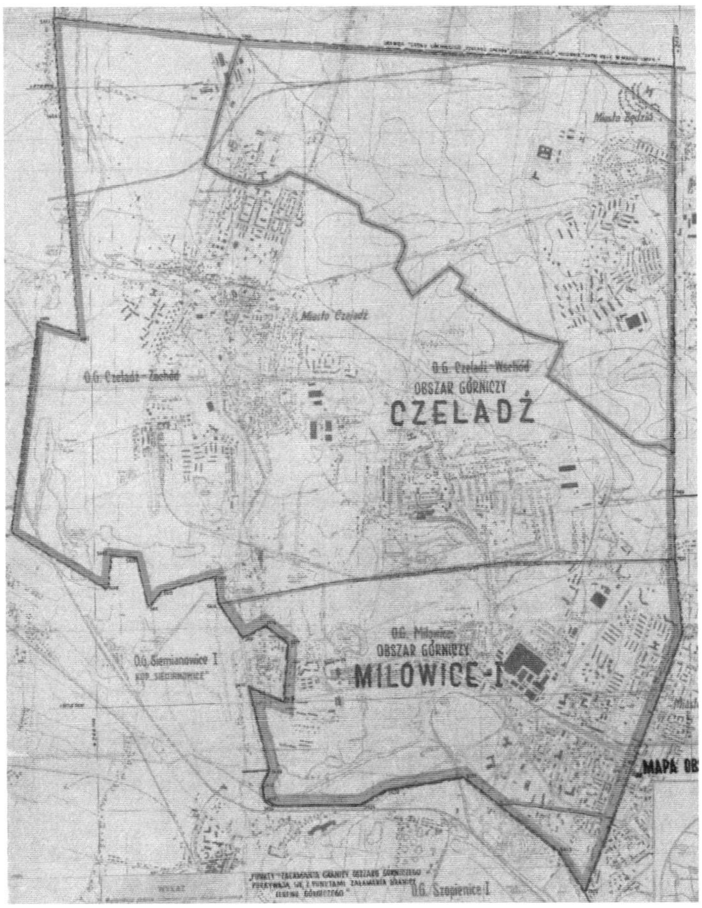

Figure 1. A fragment of the surface map with the boundary of the "Saturn" mining area marked out (materials obtained from the Archive of Mine Survey and Geological Documentation of the State Mining Authority in Katowice).

The limit of the mining area of the "Saturn" mine was 31 points. The coordinate system in which the mining maps were carried out was the Sucha Góra local system (table 1). In addition, these points had coordinates converted into the 1965 system.

Table 1. Coordinates of the refraction points of the mining area of the Saturn Mine (own elaboration based on materials obtained from the Archive of Mine Survey and Geological Documentation of the State Mining Authority in Katowice).

No	Point number	Coordinate system „Sucha Góra"		No	Point number	Coordinate system „Sucha Góra"	
		X	Y			X	Y
1	5403	-8003.00	12540.00	17	5419	-13484.20	14033.70
2	5404	-8774.00	12590.00	18	5420	-14176.50	14193.70
3	5405	-10592.00	12708.00	19	5433	-14287.30	14131.10
4	5406	-10480.00	12372.00	20	5434	-14400.00	14138.00
5	5407	-12193.29	12100.90	21	5435	-14457.00	14232.50
6	5408	-12664.36	12906.28	22	5436	-14468.70	14950.00
7	5409	-12393.15	12924.06	23	5450	-14350.00	15190.00
8	5410	-12409.53	13158.84	24	5451	-14259.00	15844.00
9	5411	-12822.19	13244.70	25	5452	-14871.00	16641.00
10	5412	-12694.66	13359.40	26	7302	-15206.00	16865.00
11	5413	-12579.50	13809.40	27	7301	-13107.00	17740.00
12	5414	-12802.81	13826.37	28	7251	-12514.00	17680.00
13	5415	-13086.00	14000.00	29	7129	-12000.00	17677.00
14	5416	-13000.00	14220.00	30	7053	-8524.00	17751.00
15	5417	-13140.00	14400.00	31	7054	-8238.00	14111.58
16	5418	-13601.70	14334.00				

"Saturn" coal mine was put into liquidation by the decision of the Minister of Industry and Trade of October 5th, 1992, No. 168/92 / Org. In the area of the "Saturn" mine there were over thirty mine workings connected to the surface (mainly shafts). The shafts were deepened from 1855. Their liquidation was carried out at different times. The name of shafts and the year of their exploration and liquidation is presented in table 2.

3. Source of materials

According to the current regulations [15, 16] it is the responsibility of every entrepreneur to keep and update geological and metrological documentation during mining operations and after prior arrangement of this documentation, forwarding it to the President of the State Mining Authority. The ordering and completeness of the surveying and geological documentation is monitored by a mining surveyor or a mining geologist under the supervision of a competent local mining supervision authority [16]. The surveying and geological documentation includes documents: measuring, computational and cartographic presenting the geological and mining situation of the mining plant, as well as the state of the area within the boundaries of the mining area [15]. The types of documents included in the survey documentation are specified in detail in the regulation [16]. One of such documents is a mining map. According to the standard [17], mining map is a cartographic document prepared by authorized person, presenting the situation and relief, mining excavations situation, geological situation, prepared by geometric projections or a mapping method, intended for mining operations. This document is to ensure proper and safe operation of mining plant operations [17].

Table 2. List of liquidated shafts in the area of "Saturn" mine (own elaboration based on materials obtained from the Archive of Mine Survey and Geological Documentation of the State Mining Authority in Katowice).

Shaft's name	Year of the exploration	Year of the liquidation
Wentylacyjny Nr 1	about 1900	no data
Rozalia	before 1905	no data
Wentylacyjny II	before 1910	about 1920
Wentylacyjny III	before 1905	about 1920
Milowice	before 1919	no data
„15”	no data	no data
Aleksander R-I	no data	no data
„16”	before 1901	1958
Antoni	before 1918	1958
XIII	no data	1963
Feliks	no data	1956
Nr 3	no data	1966
Korneliusz	no data	1966
Alfred	1908	1975
Wiktor	1898	1989
VI	1956	1980
Jan	1908	1980
Łoboda	1900	1987
Wojciech	1890	1986
Julian	1900	1989
Pogoń-C	1954	1993
V	no data	1994
Anna	1855	1994
N-II	1966	1993
N-I	1987	1993
Abraham	before 1900	1995
Kondratowicz	1924	1995
Piotr	before 1900	1996
IV	1927	1996
Aleksander R-II	1880	1996
Pogoń-M	1957	1996
Hieronim	before 1900	1996

In the case of "Saturn" coal mine, after making the decision on its liquidation, the geological and metrological documentation (including cartographic documentation) went to the Archive of Mine Survey and Geological Documentation of the State Mining Authority in Katowice), from where the author obtained a map of the mining area of the mine. On its basis, it determined the location of liquidated mine workings connected to the surface. The obtained map was in the form of a raster image and therefore it was necessary to give it a spatial orientation (the calibration process was carried out).

The software used (QGIS program) allows to calibrate the map based on the coordinates of characteristic points or by indicating such points on a previously calibrated map, or on a layer shared in free services (Google maps, OpenStretMap, etc.).

In the case of a map of the surface of the "Saturn" mining area, the author calibrated it in two different ways. In the first of them the coordinates of selected breakpoints of the mining area were used, in the second one a map from Google Maps was used. In both cases, the transformation was carried out using the Helmert method.

As already mentioned, the map of the mining area of the "Saturn" mine was conducted in the Sucha Góra local system. In addition, each of the points had specified coordinates in the 1965 layout. Due to the fact that the QGIS program does not allow the calibration of maps in local systems, coordinates in the 1965 system were used. Calibration was carried out based on several selected points (ten points were used in total the border of the mining area, evenly distributed over the entire map sheet). A fragment of the calibrated surface map is shown in figure 2.

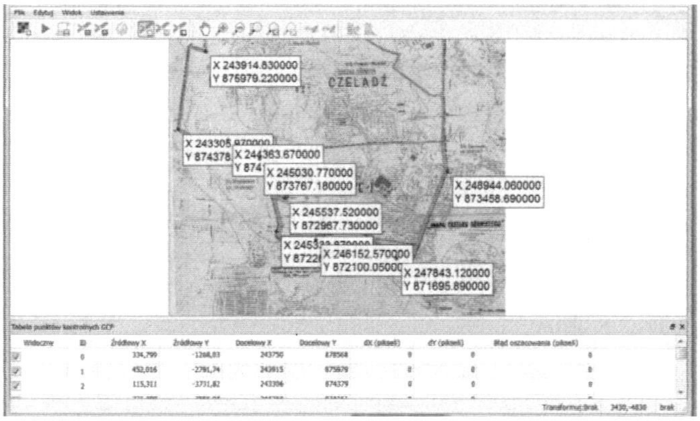

Figure 2. Fragment of the calibrated surface map.

The transformation parameters are presented in table 3.

Table 3. Parameters of transformation for points with known coordinates in the system 1965.

Shift x	Shift y	Scale x	Scale y	Rotation (degrees)	Medium error (map units)
243227.915	880737.730	1.70004	1.70004	-1.03322	16.4102

In the case of calibration based on map points obtained from the Google Maps service, it was necessary to identify them earlier on both layers (calibrated map and the layer loaded into QGIS from a given site). The fragment of the surface map and the corresponding area on the layer from Google Maps are shown in figure 3.

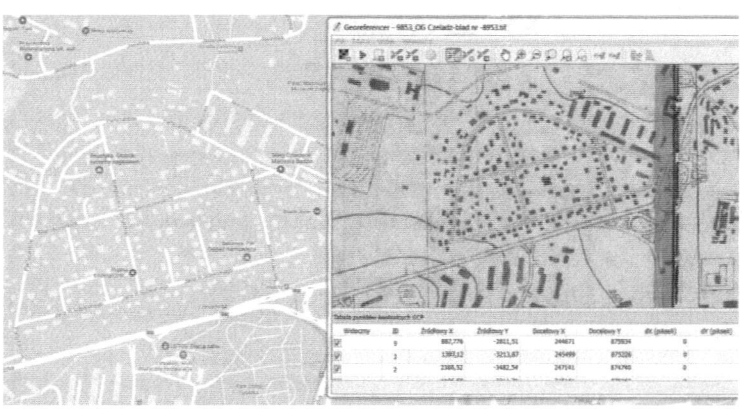

Figure 3. Selection of points on the basis of which calibration will be carried out (area identification).

Also in this case, the calibration was carried out on the basis of ten points evenly distributed on the map sheet. The fragment of the map of the "Saturn" mine surface with the points included in the calibration is shown in figure 4.

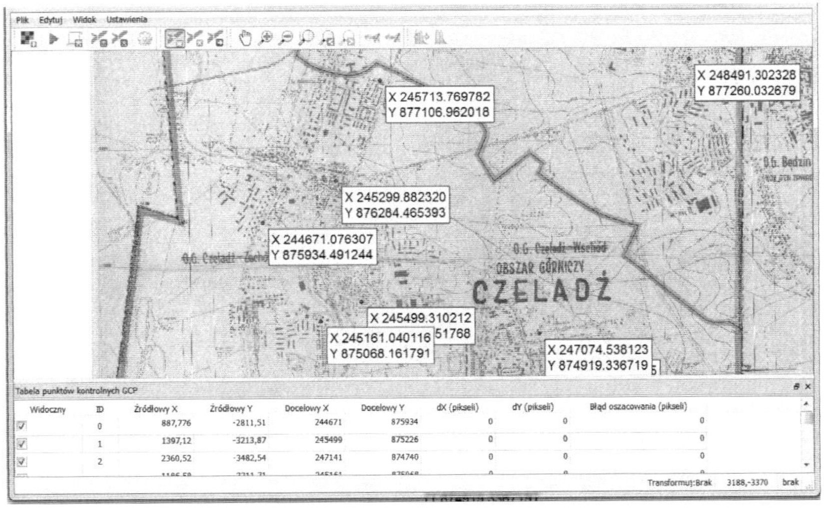

Figure 4. A fragment of the map of the surface of "Saturn" mine with the points included in the calibration.

The transformation parameters are presented in table 4.

Table 4. Parameters of transformation for points located inside the border of the mining area.

Shift x	Shift y	Scale x	Scale y	Rotation (degrees)	Medium error (map units)
243236.730	880751.465	1.70473	1.70473	-1.13705	16.0016

An example of the shafts' location against the current spatial situation is shown in figure 5 and figure 6.

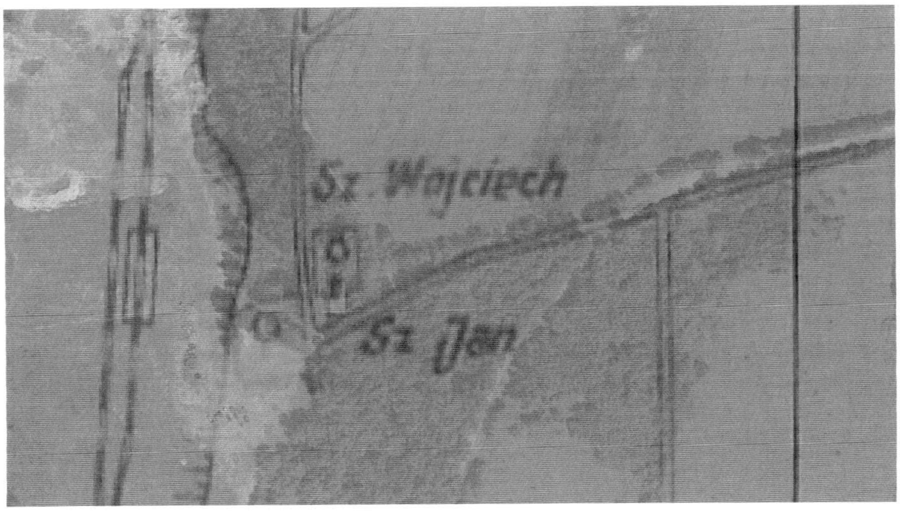

Figure 5. Location of the Wojciech i Jan shaft on the background of the Google Maps map.

Figure 6. Location of the Pogoń-M shaft on the background of the map from Google Maps.

In both cases, the average calibration error obtained was at a similar level (around 16m). Considering the scale of the map and its quality, the obtained result should be considered satisfactory.

After giving the spatial reference map, it is possible to observe changes that occurred in the period from the liquidation of a excavation to the present day. For example, the effect of calibration is shown in Figure 8 and figure 9. In the case of Jan and Wojciech shafts, the changes were small (figure 8). In the case of the area on which the shaft Pogoń-M liquidated in 1996 is located, the changes were much larger. A truck car park is located above the dismantled shaft.

4. Threats related to the presence of abandoned shafts

Liquidation of mining excavations, having connection to the surface, can be carried out in several ways. One of them is cutting off the shaft with insulating stoppings on the pit bottom and filling the shaft pipe with a suitable material. The inlet of the shaft is additionally covered with a concrete or reinforced concrete slab on which the embankment is made. According to the authors of the report [11], this is the most appropriate and the most secure method. Another way is to secure the shaft tube with a plate or a platform on which the embankment is made. Liquidation of the shaft is also possible by damming inlets to the pit bottoms and covering the shaft pipe.

According to the information in table 2, the liquidation of shafts in the area of "Saturn" mine has been carried out over seventy years. In the case of five shafts (Wentylacyjny Nr 1, Rozalia, Milowice, 15, Aleksander R-I) there is no information of the time of their liquidation. Very often documentation describing the manner of liquidation of such excavations has been preserved in a form that is insufficient to unambiguously determine whether it was carried out in a correct manner and does not threaten the use of these areas in later times.

After the shafts liquidation for their condition and their surroundings, the various processes taking place in the rock mass are inter alia: water circulation, corrosion, suffosion, embankment compaction, etc. [18, 19]. As a consequence, discontinuous deformations may occur, most often in the form of depressions [13, 20, 21]. As indicated by the authors of the works [12, 13, 14, 21], liquidated mine workings connected with the surface may also pose a threat to the atmospheric air, mainly due to the possibility of outflow from high amounts of gases such as methane or carbon dioxide. Reliable information about the location of this type of workings would allow to increase safety in their area.

5. Summary

The article discusses the issues related to the use of modern IT tools in the mining industry. Discussed is the method of obtaining information on the location of liquidated mining excavations, having connection to the surface, based on archival maps, using the GIS class software. Despite the rapid development of measurement techniques (eg GNSS, remote sensing, photogrammetry, etc.), traditional maps are still an important source of information, in particular for geographic information systems. Using the QGIS program, the author calibrated the map of the surface of the mining area of the liquidated „Saturn" coal mine. The prepared raster layer made it possible to reconstruct the location of shafts located in this area. The excavations were subjected to gradual liquidation from the beginning of the 19th century (eg Ventilation shaft II and Ventilation shaft III). For some of them there is no information about the year of liquidation. Very often there are no documents specifying their location (measurement documents) or they are dispersed in various institutions. In addition, documents constituting measurement and geological documentation in many cases (especially from the initial period of activity of a given mine) have been preserved in residual and incomplete form.

Locating places where excavations of this type had a connection with the surface is important because their presence is a threat due to the possibility of discontinuous deformations and the outflow of methane or carbon dioxide.

Knowledge about this subject will allow to label properly these places and keep up to date with the state of the excavations as well as their surroundings.

The author of the article showed that the QGIS program, belonging to the GIS class programs, easily allows obtaining information from archival maps, as well as has the ability to develop a database of this type of objects.

References

[1] Gaździcki J 1990 *Systemy informacji przestrzennej* (Warszawa: Państwowe Przedsiębiorstwo Wydawnictw kartograficznych)

[2] Sokoła-Szewioła V and Żogała M 2016 The adoption of the ArcGIS system to support the analyses of the influence of the mining tremors on the building objects *Management Systems in Production Engineering* **4 (24)** pp 228 – 36

[3] Sokoła-Szewioła V, Poniewiera M and Staniek I 2017 Opracowanie bazy danych o obiektach budowlanych posadowionych na terenie górniczym z wykorzystaniem wolnego oprogramowania *Przegląd Górniczy* **12** pp 47 – 54

[4] Klemens J and Poniewiera M 2010 Wykorzystanie najnowszych technologii informatycznych do wsparcia procesów mierniczo-geologicznych w Kompani Węglowej S.A. *Przegląd Górniczy* **10** pp 3-10

[5] Blachowski J and Nowacka A 2011 Internetowy system geoinformacyjny o dawnych obiektach górniczych na przykładzie kopalni węgla kamiennego „Julia" w Wałbrzychu *Prace Naukowe Górnictwa Politechniki Wrocławskiej Studia i Materiały* **133** pp 35 – 47

[6] Kura K 2007 Budowa i funkcjonowanie systemu informacji o terenie oraz przykłady zastosowań dla obszarów kopalń węgla kamiennego w GZW *Prace naukowe GIG Górnictwo i Środowisko* **8** pp 53 – 63

[7] Świtoń J 2013 Budowa bazy danych graficznych i opisowych do trójwymiarowego modelu kopalni i pokładów węgla w środowisku ArcGIS *CUPRUM Czasopismo Naukowo-Techniczne Górnictwa Rud* **2** pp 37 – 48

[8] Iwańczak B 2016 *QGIS Tworzenie i analiza map* (Gliwice: Helion)

[9] Nowotarska M 2009 *Wprowadzenie do Quantum GIS* Szczecin-Wrocław

[10] Szczepanek R and Zmuda-Trzebiatowski P 2017 *Systemy informacji przestrzennej z QGIS* http://www.dts.put.poznan.pl/samouczek-qgis/

[11] Goszcz A at al. 1996 *Wykonanie badań górotworu w obrębie byłych szybów i upadowych w celu określenia ich wpływu na zagrożenie powierzchni a w razie potrzeby opracowanie optymalnego sposobu likwidacji tych wyrobisk* Przedsiębiorstwo Gemes Sp. z o.o.

[12] Wrona P, Różański Z and Pach G 2017 Emisja gazów cieplarnianych z nieczynnego szybu – uwarunkowania, ocena i profilaktyka *Power & Industry* **1** pp 50 – 2

[13] Czaja P and Gąsiorczyk A 2005 Wpływ sposobu likwidacji szybu na zagrożenie powierzchni w jego otoczeniu *Górnictwo i Geoinżynieria* **3/1** pp 165 – 75

[14] Mitka J, Wątor L, Łukomski R, Wita A, Birawski A, Gerwin M and Musiał E 2015 Technologia pozyskiwania metanu powierzchniowymi otworami wiertniczymi z wyrobisk zlikwidowanych kopalń węgla kamiennego dla gospodarczego wykorzystania i ochrony środowiska terenów górniczych *XX Konferencja Naukowo-Techniczna* Rybnik
[15] Ustawa Prawo geologiczne i górnicze z dnia 9 czerwca 2011 r. Dz.U. 2017 poz. 2126
[16] Rozporządzenie Ministra Środowiska z dnia 28 października 2015 r. w sprawie dokumentacji mierniczo-geologicznej Dz.U. 2015 poz. 1941
[17] PN-G-09000-1. 2002 Mapy górnicze Część 1: Podział i terminologia
[18] Strzałkowski P 2007 *Ochrona środowiska na terenach górniczych. Wybrane problemy* Monografia (Gliwice: Wydawnictwo Politechniki Śląskiej)
[19] Popiołek E 2009 *Ochrona terenów górniczych* (Kraków: Wydawnictwo AGH)
[20] Czaja P 2009 Ocena rozwiązań projektowych likwidacji szybów zastosowanych w procesie restrukturyzacji polskiego górnictwa węglowego *Górnictwo i Geoinżynieria* **3/1** pp 105 – 19
[21] Wojtacha P, Orlof J and Picur J 2015 Ocena zagrożeń ze strony wyrobisk górniczych mających połączenie z powierzchnią usytuowanych w zlikwidowanych podziemnych zakładach górniczych *Bezpieczeństwo pracy i ochrona środowiska w górnictwie* **2** pp 3 – 11

IOP Conf. Series: Earth and Environmental Science **174** (2018) 012010 doi:10.1088/1755-1315/174/1/012010

Searching for dependent air currents in a mine ventilation network using the connected graph disconnection method

Grzegorz Pach

Faculty of Mining and Geology, Silesian University of Technology, Akademicka 2, 44-100 Gliwice, Poland

grzegorz.pach@polsl.pl

Abstract. Ventilation networks in underground hard coal and copper ore mines are usually designed with branches with dependent air currents. Among these, branches connecting sub-networks of main fans may be distinguished. In view of ventilation hazards: fire hazard, gas hazard, thermal hazard and coal dust explosion hazard, the presence of dependent air-currents may be both desirable and adverse. Information regarding the possible presence and character of the dependent air currents in the mining plant's ventilation network is thus significant.
The paper presents a method, based on graph theory, allowing to determine the presence of dependent air currents and their affiliation with the supplied (fresh) air or the removed (used) air. This allows to obtain information whether air currents connecting the main fan sub-networks exist in a ventilation network. The method is applicable in all ventilation networks under the condition of the previous reduction of parallel connections.
The paper presents an example of a ventilation network with three main fans, where the method in concern was applied. The algorithm presented in the article allowed to determine the existence of dependent currents in each of the air zones. Additionally, it allowed to identify the main ventilator sub-networks that are interconnected.

1. Introduction

The Polish hard coal mining industry is continuously subject to restructuring. As a part of this process, not only unprofitable mines are being shut down, but also mining plants are being merged [1]. The effects of such undertakings include significant changes in the structure of the mines' ventilation networks. Differences occurring in the structure of these networks are significant in the management of the ventilation of an underground mining plant [2], especially in the analysis and prevention of ventilation hazards: fire hazard, methane hazard, thermal hazard and coal dust explosion hazard [3]. The structure of an underground mining plant's ventilation system also impacts the distribution of smoke and gas produced in fire in headings, being thus decisive for the area of the zone with smoke conditions or the applicability of reverse ventilation [2, 3, 4]. In case of methane hazard, methane removal efficiency and the supply of adequate amounts of air to the exploitation areas is highly dependent on the size and the complexity of the ventilation network [5]. One of the preventive methods applied in cases of thermal hazard is to ensure an increased air flow through workings [6, 7]. The possibility to reach an increased airflow is dependent on the structure of the ventilation network.

In cases where mining plants are merged, the newly formed ventilation network may and usually does exhibit a complex structure. In such networks, dependent air currents may occur both in the supplied (fresh) and the removed (used) air zones. Among these, currents which connect the sub-networks of the main ventilation fans may appear. Due to the ventilation hazards in mining plants, information regarding the presence of dependent air currents and the type of the currents in the ventilation network is highly significant.

The purpose of this paper is to present a method allowing for the determination whether dependent air currents occur in the analyzed underground mining plant as well as to determine the type of the currents. The presented method is applicable for ventilation networks with all structures.

2. Types of air currents in mining plant ventilation networks

Workings of underground mining plants are interconnected, thus forming a mine ventilation network. The network includes: nodes, branches, fans and other ventilation devices (e.g. various types of dams) [8]. A division is known (due to the number of incoming and outgoing air streams) into the following types of nodes: separating, combining and mixing (figures 1a-1c) [9]. In mining practice, the application of the so-called semi-node is also known (figure 1d).

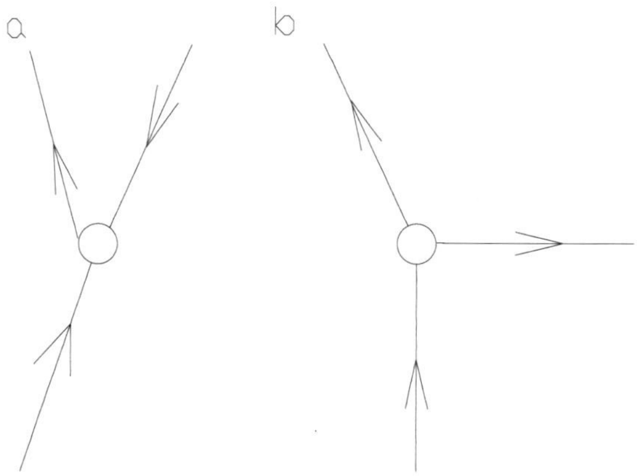

Figure 1. Types of nodes in a mine ventilation network: a) combining node b) separating node c) mixing node d) semi-node.

A working or a series of workings connecting two nodes is called the ventilation branch. The air current flowing in a branch may be normal (with single direction of flow) or diagonal (the direction depends on the aerodynamic resistance in neighbouring branches). A. Krach [10] has presented one of the methods to determine the diagonal character of a branch.

Branches through which the air flows to the so-called take-off points (e.g. longwalls, functional chambers) are called branches with supplied air (fresh), while branches through which the air flows out from the take-off points up to the surface are called the branches with removed air (used). Due to the above, all branches in a mine ventilation network may be divided into:

- branches with supplied air currents (fresh),
- branches with air take-off points,
- branches with removed air currents (used),
- branches representing the atmosphere.

All branches with supplied air current form a zone of supplied (fresh) air while the branches with the removed air constitute a removed (used) air zone. A similar division may also be applied to the nodes of the ventilation network. After distinguishing such zones, another division of air currents may be made:

- independent air currents – separating from the supplied air current, ventilating the branches with take-off points and joining the removed air current,
- dependent air currents – connecting two different supplied air currents or connecting two different removed air currents.

Figure 2 presents a canonical diagram of the mine ventilation network. The green p-c line denotes the full section through the ventilation network, passing through all branches with take-off points (4 mining regions and 1 functional chamber). The section divides the ventilation network into two zones: the supplied air zone (branches with red-coloured arrows) and the removed air zone (branches with blue-coloured arrows). The broken line connecting the nodes 1-12, 9-12 and 11-12 represents branches symbolizing the earth's atmosphere. While analyzing the above canonical diagram, it may be concluded that the air current in the 7-13 branch (based on the node numbers) is diagonal, while the current in the 1-2 branch is normal. The air current in the 22-7 branch is an independent current, ventilating the mining region, while the air current in the 12-8 branch is a dependent current. Due to the location in given zones, the air current in the branch 4-6 is a dependent air current in the supplied air zone, while the air current in the branch 12-10 is a current in the removed air zone. Additionally, air current in the branch 12-10 connects the sub-networks of the main ventilation fans. It should be noted that each dependent air current is a diagonal current.

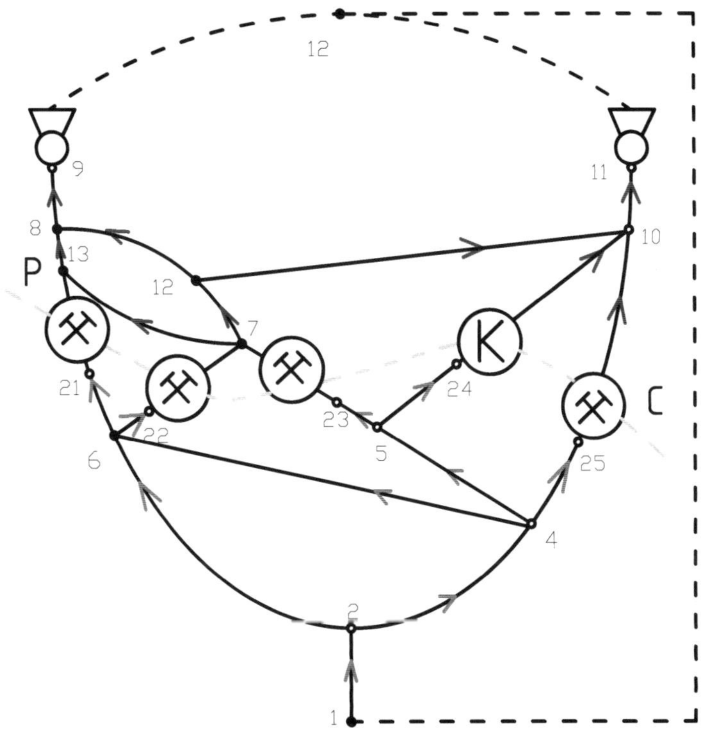

Figure 2. Example of a ventilation network with dependent currents in the supplied (fresh) and removed (used) air zones.

3. Advantages and disadvantages of the occurrence of dependent currents in ventilation networks of mines

In mining practice, ventilation networks containing only branches with normal air currents are very rare. In the Polish hard coal and copper ore mining industry, ventilation networks almost always include branches with a diagonal character and dependent air currents.

The advantages of the existence of dependent air currents include:

- shortening of escape routes,
- increased possibilities of establishing transport routes for the mined material, staff and materials,
- the possibility to decrease the main ventilation costs [11].

In case of a fire in the mining region in the 25-10 branch (figure 2) due to the existence of the 12-10 dependent current, it is sufficient for an employee of that division to move to node 10, where air, which does not contain smoke, flows from node 12.

On the other hand, the disadvantages of the dependent currents are:

- the possible increase of the area of the direct smoke zone,
- difficulties in the design of mining plant ventilation systems,
- increase of the costs related to the maintenance of additional workings.

In case of a fire in the branch 2-4, due to the presence of the branch with dependent current 4-6, all mining regions as well as the functional chamber are affected by smoke and not only the regions 23-7, 25-10 and the 24-10 functional chamber.

4. Method for searching for dependent air currents

Solutions applying the graph theory for solving problems related to a mining plant's ventilation system are known in source literature. Such activities are substantiated and allow to apply mathematical means offered by the theory [12]. The theory finds its application also in the problem in concern.

The proposed method allows to answer the following questions:

- Are there branches with dependent currents in the investigated ventilation network of the underground mining plant?
- In which air zone are the branches located?
- Does the ventilation network contain branches connecting sub-networks of the main ventilation fans?

The liquidation of parallel connections in a ventilation network is a prerequisite to apply the method. The presented method consists in the disconnection (disintegration) of a connected graph representing a ventilation network into independent sub-graphs. The determination of the cyclomatic number [12] for each of the sub-graphs makes it possible to answer the above questions.

The algorithm of the method consists in a series of subsequent operations.

1. Removal of branches with air take-offs (or, generally, branches contained within the full section passing through the air take-offs) and branches representing the atmosphere from the structural representation of the network. Such an operation will disjoin the graph representing the ventilation network into several disconnected subgraphs.
2. Assignment of nodes and remaining branches to individual subgraphs and adding them up.
3. The determination of the cyclomatic number value v for each of the subgraphs based on the dependence 1.

$$v = m - n + 1 \tag{1}$$

where: m – number of branches in a zone, n – number of nodes in a zone.

Interpretation of results of the algorithm.

1. If the cyclomatic number in the supplied/removed air zone is more than 0, this indicates the existence of dependent current in these zones. The 0 value denotes that no such currents exist in the tested network.
2. Assuming that a *lw* number of main fans exist in a ventilation network, the following cases are possible after the disconnection of the graph representing the ventilation network:
   a) Two subgraphs were formed, one for the supplied air zone and the second for the removed air zone. This indicates the interconnection of all main fan subnetworks with dependent currents.
   b) *lw*+1 subgraphs were formed. It may then be concluded that no currents connecting any of the main fan subnetworks occur.
   c) At least 3, but not more than *lw* subgraphs were formed. In relation to the above, this is an intermediate situation – some but not all subnetworks of the main fans are connected.

5. Example of searching for dependent currents

The example of the operation of the method shall be presented with the ventilation network presented in figure 3. In the analyzed network, 6 branches with take-off points exist (4 mining regions –

branches 6-8, 6-7, 5-7 and 4-10 as well as two functional chambers – branches 5-13, 5-10). Branches representing the atmosphere are the branches: 9-12, 14-12, 11-12 and 12-1. In line with the first step of the algorithm, all the mentioned branches shall be removed. After the removal of these branches, a transition matrix representing the ventilation network is developed – presented in table 1.

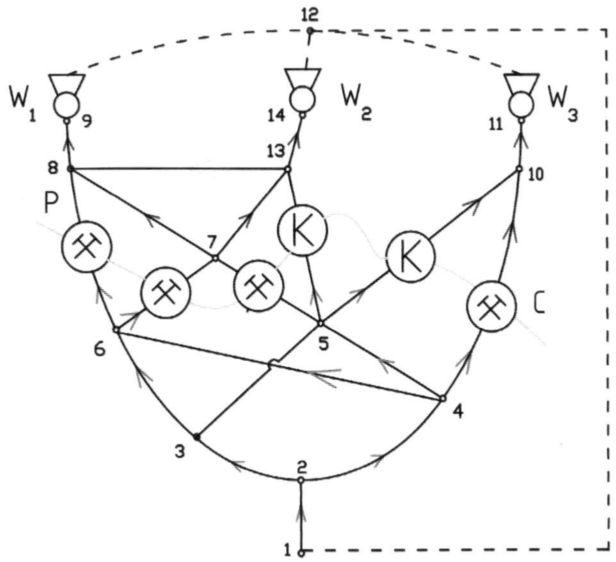

Figure 3. Canonical diagram of the A mine ventilation network.

Table 1. Transition matrix for the A mine ventilation network.

w_j / w_i	**1**	**2**	**3**	**4**	**5**	**6**	**7**	**8**	**9**	**10**	**11**	**12**	**13**	**14**
1	0	1	0	0	0	0	0	0	0	0	0	*0*	0	0
2	-1	0	1	1	0	0	0	0	0	0	0	0	0	0
3	0	-1	0	0	1	1	0	0	0	0	0	0	0	0
4	0	-1	0	0	1	1	0	0	0	**0**	0	0	0	0
5	0	0	-1	-1	0	0	**0**	0	0	**0**	0	0	**0**	0
6	0	0	-1	-1	0	0	**0**	**0**	0	0	0	0	0	0
7	0	0	0	0	**0**	**0**	0	1	0	0	0	0	1	0
8	0	0	0	0	0	**0**	-1	0	1	0	0	0	-1	0
9	0	0	0	0	0	0	0	-1	0	0	0	*0*	0	0
10	0	0	0	**0**	**0**	0	0	0	0	0	1	0	0	0
11	0	0	0	0	0	0	0	0	0	-1	0	*0*	0	0
12	*0*	0	0	0	0	0	0	0	*0*	0	*0*	0	0	*0*
13	0	0	0	0	**0**	0	-1	1	0	0	0	0	0	1
14	0	0	0	0	0	0	0	0	0	0	0	*0*	-1	0

Bold font indicated the zeroed cells of the transition matrix, corresponding to the removed branches with air take-off points, while cursive denotes the zeroed cells corresponding to the removed branches representing the atmosphere.

Subsequently, in line with the second step of the algorithm, the first subgraph is formed, in which node 1 is included. While analyzing the first row of the matrix (table 1), one may note that node 1 is connected to node 2. Node 2 is thus included in the first subgraph. Subsequently, the row of the matrix characteristic to the node 2 is considered. Here, it is found that the node 2 is connected to nodes 3 and 4. Thus, the nodes 3 and 4 are included in the first subgraph. Subsequently, the rows of the matrix corresponding to nodes 3 and 4 are analyzed and nodes 5 and 6 are added to the first subgraph. During the analysis of the rows corresponding to nodes 5 and 6, no new nodes may be found that could be attributed to the first subgraph. The first subgraph shall thus contain the nodes: 1, 2, 3, 4, 5 and 6. After another analysis of the matrix, it may be found that the following branches are related to the nodes: 1-2, 2-3, 2-4, 3-5, 3-6, 4-5 and 4-6. The number of nodes in the first subgraph shall thus amount to 6 while the number of branches shall be 7.

Subsequently, a second subgraph is formed starting with the next "free" node (in this case, node 7) and the procedure presented above is repeated. The second subgraph includes the nodes and branches7-8, 7-13, 8-9, 8-13 and 13-14. The number of nodes and graphs in the second subgraph is identical and amounts to 5.

Next, a third subgraph is formed. This subgraph includes nodes 10 and 11 and the 10-11 branch.

Node 12 may not be included in any of the subgraphs as it is not connected to any other node (table 1).

Based on the formula 1, values of the cyclomatic number are determined for each of the subgraphs. These amount to: for the first subgraph – 2, for the second subgraph – 1, for the third subgraph – 0.

5.1. Analysis of the obtained results

The cyclomatic number of the first subgraph is 2 and, simultaneously the node 1 (intake to the shaft) is in the supplied air zone – thus dependent currents exist in this zone.

The cyclomatic number of the second subgraph is 1 and nodes 9 and 14 are connected with branches to fans W1 and W2 (figure 3), thus dependent air currents also occur in the removed air zone.

The number of the formed subgraphs is 3, while the number of the main fans is also 3. Thus a situation presented in subsection c (chapter 4) occurs. Some of the main fan subnetworks are interconnected. The second subgraph includes nodes 9 and 14 (nodes with fans W1 and W2), so the subnetworks of these two fans are interconnected.

To confirm the analysis of the results, figure 4 presents the subgraphs that were formed after the removal of branches with air take-off points and branches representing the atmosphere.

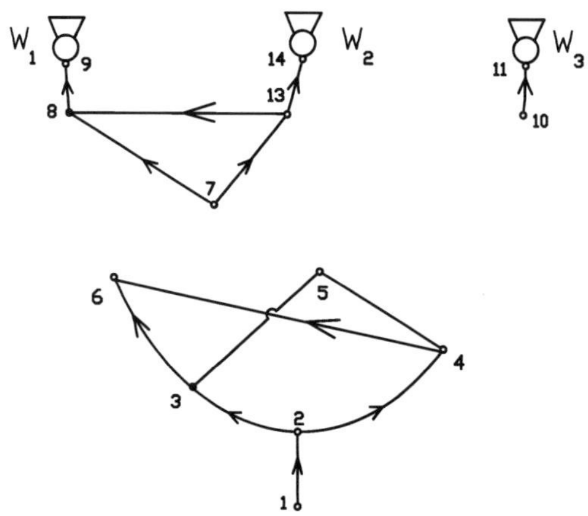

Figure 4. Subgraphs that were formed after the disconnection of the graph representing the A mine ventilation network.

6. Conclusions

1. The method using elements of the graph theory presented in the paper is applicable to any mine ventilation network, under the condition that the parallel connections of branches are previously reduced.
2. The method allows to determine whether dependent currents occur in a mine ventilation network and to specify their location in the supplied (fresh) and removed (used) air zones.

IOP Conf. Series: Earth and Environmental Science 174 (2018) 012010 doi:10.1088/1755-1315/174/1/012010

3. By comparing the number of subgraphs formed after the disconnection of the graph representing the ventilation network of the mine with the number of main fans, information is obtained regarding the existence or non-existence of a connection between the main fan subnetworks.
4. The information regarding the presence of dependent air currents, including currents which connect subnetworks of main fans is significant in the prevention practices applied for ventilation hazards. The presence of such currents may be both beneficial and disadvantageous.

References

[1] Kugiel M 2010 Efekty procesów restrukturyzacyjnych w Kompanii Węglowej S.A. *Kwartalnik Górnictwo i Geologia* **5(3)** pp 59–73

[2] Wacławik J 2010 *Wentylacja kopalń - tom 1* (Kraków: Wydawnictwa AGH) pp 282–318

[3] Szlązak N and Zając K 1998 *Ocena możliwości wykonania rewersji wentylacji głównej w kopalniach węgla kamiennego (Kraków:* Biblioteka Szkoły Eksploatacji Podziemnej)

[4] Strumiński A 1996 *Zwalczanie pożarów w kopalniach głębinowych (Katowice:* Wydawnictwo „Slask")

[5] Krause E and Łukowicz K 2012 Wpływ struktury kopalnianej sieci wentylacyjnej na skuteczność ujęcia metanu *Prace naukowe Głównego Instytutu Górnictwa, Kwartalnik Górnictwo i Środowisko* **4** pp 95–108

[6] Drenda J, Sułkowski J, Różański Z, Pach G and Wrona P 2016 Two stage assessment of thermal hazard in an underground mine *Archives of Mining Science* **61** pp 309–22

[7] Drenda J, Pach G, Różański Z, Wrona P and Sułkowski J 2018 Safe working conditions in hot mine environment – the analysis of different indices *Archives of Mining Science* **63** pp 93–106

[8] Musioł D and Pach G 2014 *Projektowanie rozpływu powietrza w sieciach wentylacyjnych przykłady obliczeniowe* (Gliwice: Wydawnictwo Politechniki Śląskiej)

[9] Kolarczyk M 1993 Wpływ struktury kopalnianej sieci wentylacyjnej na wrażliwości prądów powietrza przy zmianach oporów bocznic *Zeszyty Naukowe Politechniki Śląskiej* **214**

[10] Krach A 2014 Determining diagonal branches In mine ventilation networks *Archives of Mining Science* **59** pp 1097–105

[11] Pach G, Sułkowski J, Różański Z and Wrona P 2018 Costs reduction of main fans operation according to safety ventilation in mines - a case study *Archives of Mining Science* **63** pp 43–60

[12] Wilson R 2017 *Wprowadzenie do teorii grafów* (Warszawa: Wydawnictwo Naukowe PWN)

IOP Conf. Series: Earth and Environmental Science **174** (2018) 012011 doi:10.1088/1755-1315/174/1/012011

Pollution of coals of the Upper Silesian Coal Basin from the perspective of the European registers of the release and transfer of pollutants to air - the case of chlorine

Szymon Pluta[1] and Franciszek Plewa[2]

[1] Logstor Polska sp. z o.o. Handlowa 1, 41-807 Zabrze, Poland
[2] Faculty of Mining and Geology, Silesian University of Technology, Akademicka 2, 44-100 Gliwice, Poland

franciszek.plewa@polsl.pl

Abstract: The regulation WE 166 (2006) of the European Parliament and European Council on the establishment of the European Register of Release and Transfer of Pollutants presents the limiting values of pollutants to the atmosphere and values from mining operations and power plants. Chlorine and its compounds constitutes a significant pollutant. Taking into account the chlorine content of selected coals (up to 2.69% in the ashes) and the limiting values of the releases into the atmosphere one should perform a detailed analysis of its release from coal mines and power plants to the environment.

1. Introduction

Hard coal is mostly an organic substance comprised of carbon, hydrogen, oxygen, nitrogen, sulphur and insignificant amounts of phosphorus. Besides the dominating content of organic substance, coal also includes mineral substance, that is, inorganic impurities found after incineration in the form of ash [e.g. 1]. The mineral components found in coal may be syngenetic – formed simultaneously to the coal substance or epigenetic – accumulated after the coal seam was formed. It has been determined that trace elements occur in the inorganic coal substance. Their type and content varies significantly. In some coals, certain elements, such as: Ge, As, B, Be, exhibit concentrations multiple times higher than in the earth crust [e.g. 2,3,4]. Concentrations exceeding the crustal abundance are also exhibited at times by: Ba, Bi, Co, Cu, Ga, La, Li, Pb, Hg, Mo, Ni, Sc, Se, Ag, Sn, V, Y and Sr. These elements include ones that constitute environmental pollutants. Threshold values for their release to air, water and soil were determined in the EC Regulation No. 166 of the European Parliament and Council dated January 18th 2006, regarding the establishment of a "European Pollutant Transfer and Release Register." Based upon this regulation, the entities releasing the substances to environment, including mining plants, CHP plants and other coal combustion systems, shall be required to control and to follow a series of requirements regarding the release of pollutants. Considering the above information, this publication presents the components of coal that are pollutants. Using the example of chlorine, its identification in the coal mines of the Upper Silesian Coal Basin was described while giving consideration to the threshold values of release to air and water. Chlorine was also selected due to the negative impact of that element on the machines used in coal processing by the formation of tarnish (chlorine corrosion) on heating devices. Moreover, chlorine is an undesirable component of coking coal as 40-60% of chlorine remains in the coke [5].

2. Components of coals in view of the "European pollutant transfer and release register" to air

Table 1 presents the components included in the coals of the Upper Silesian Coal Basin that simultaneously constitute pollutants, the release of which to air, water and soil is limited based on the European Pollutant Transfer and Release Register (Regulation No. 166/2006 of the European Parliament and Council).

Table 1. Threshold values of release of selected components of coal to water and air.

No.	Pollutants	Threshold value for release to air, kgs/year
1	Carbon monoxide	500000
2	Carbon dioxide	100000000
3	Nitrogen oxides (NO_x/NO_2)	150000
4	Total nitrogen	
5	Total phosphorus	
6	Arsenic and arsenic compounds	20
7	Cadmium and cadmium compounds	10
8	Chromium and chromium compounds	100
9	Copper and its compounds	100
10	Mercury and its compounds	10
11	Nickel and its compounds	50
12	Lead and its compounds	200
13	Zinc and its compounds	200
14	Chlorium (as total Cl)	
15	Chlorium and its inorganic compounds (as HCl)	10000
16	Fluorides (as total F)	
17	Fluorine and its inorganic compounds (as HF)	5000
18	Suspended particulate matter (PM_{10})	50000

3. Content of chlorine in the coals of the Upper Silesian Coal Basin

Chlorine – similarly to nitrogen and sulphur – is an element that is commonly occurring in hard coal. In various coals the content is different and it depends on the conditions of formation. It was determined that the content of chlorine increases along with the depth of the deposit and the level of water salinity.

As provided by T. Mielecki [6], the mean content of chlorine in the coals of the Upper Silesian Coal Basin is 0.01% (from 0 to 0.025%) in case of saddle beds (limnic type), while in the case of paralic beds, it reaches 0.175%, usually not exceeding 0.1%. J. Kuśmierska [6] exhibited that in case of certain coals, the chlorine content reaches 0.3%. On the other hand, A. Różkowska [7], while analysing the chlorine content in coals from core samples, found that the chlorine content was from 63 to 18871 g/Mg. The highest contents of chlorine are found in the coals of the eastern and south-eastern parts of the Upper Silesian Coal Basin (4000-9000 g/Mg), while the lowest contents are found in the south-western part (400-700 g/Mg). As provided by A. Strugała [5], the chlorine content in coals varies in the range from 0.02 to 0.4%, whereas M. Kosewska and K. Wróblewska [8] have exhibited that the amount of chlorine in coking coals used as a charge is within the range from 0.02 to 0.4%. Based on the tests of coal samples from the south-western part of the Upper Silesian Coal Basin, it has been found that the content of chlorine in ash after combustion reaches 4.08%. The results of chlorine content in ash after combustion of coals from the Cracow sandstone series (KSP), mudstone series (SM), the Upper Silesian sandstone series (GSP) and the paralic series (SP), have been presented in tables from 2 to 7.

Table 2. The content of chlorine in the ashes of selected coals from the Cracow sandstone series.

Lithostratigraphic series	Coal seam	Cl, %
KSP	206	1.26
KSP	206	0.51
KSP	209	2.69

Table 3. The content of chlorine in the ashes of selected coals from the Mundstone series (the Orzesze layers).

Lithostratigraphic series	Coal seam	Cl, %
SM	308	1.12
SM	348	0.79
SM	361	0.19
SM	364/1	4.08

Table 4. The content of chlorine in the ashes of selected coals from the Mudstone series (the Ruda layers).

Lithostratigraphic series	Coal seam	Cl, %
SM	403/3	2.65
SM	404/4	0.15
SM	405/2	0.44
SM	409/4	1.68

Table 5. The content of chlorine in the ashes from selected coals of the Upper Silesian sandstone series.

Lithostratigraphic series	Coal seam	Cl, %
GSP	502/2	0.42
GSP	503	0.37
GSP	505	0.68
GSP	507	0.23
GSP	510/1	0.25

Table 6. The content of chlorine in the ashes from selected coals of the Paralic series (the Porębskie Layers).

Lithostratigraphic series	Coal seam	Cl, %
SP	600	0.91
SP	626/1	0.05

Table 7. The content of chlorine in the ashes from selected coals of the Paralic series (the Jaklowieckie layers).

Lithostratigraphic series	Coal seam	Cl, %
SP	712/2	0.68
SP	713/1	0.96

The use of hard coal, consisting mostly in its combustion, results in the release of the chlorine present in the coal, most often in the form of hydrochloride, which is a compound that is aggressive to power and processing machines. Considering this issue, the results of the chlorine content tests of coals from various lithostratigraphic series and giving consideration to the threshold value in line with the regulation No. 166/2006 of the European Parliament and Council amounting to 10000 kgs/year, the content of chlorine (hydrochloride) released to air as gas and as particulate matter during the combustion in CHP plants should be analyzed.

4. Summary

Chlorine is a common component in the coals of the Upper Silesian Coal Basin and – at the same time – is a significant pollutant. This is why the threshold values for its release to air and water have been specified in the regulation (EC) No. 166/2006 of the European Parliament and Council in the established "European Pollutant Release and Transfer Register." Tests have exhibited that the chlorine content in ash after the combustion of selected coals of the Upper Silesian Coal Basin reaches 2.69%. Considering these results, detailed analyses of conditions of disposal of chlorine from coal mines and power plants should be conducted.

References

[1] Parzentny H R 1995 *Wpływ nieorganicznej substancji mineralnej na zawartość niektórych pierwiastków śladowych w węglu Górnośląskiego Zagłębia Węglowego* Wydawnictwo Uniwersytetu Śląskiego Katowice p 90

[2] Kuhl J and Ziółkowski J 1954 Pierwiastki rzadkie w górnośląskim węglu *Przegląd Górniczy* **10** pp 180–92

[3] Widawska- Kuśmierska J 1981 Występowanie pierwiastków śladowych w polskich węglach kamiennych *Przegląd Górniczy* **7** pp 445–59

[4] Strugała A 1998 Substancja mineralna węgla kamiennego i jej przemiany w procesie koksowania *Gospodarka Surowcami Mineralnymi* **41/1**

[5] Mielecki T 1948 Przyczynek do poznania zawartości chloru w polskich węglach *Biuletyn Instytutu Naukowo-Badawczego Przemysłu Węglowego* Komunikat nr 36 pp 1–5

[6] Kuśmierska J 1984 Zanieczyszczenia polskich węgli kamiennych w aspekcie ochrony środowiska *Problemy badań węgla w pracach geologiczno-złożowych nowych technologii jego utylizacji* pp 94–104

[7] Różkowska A 1987 Zawartość chloru w węglach kamiennych Górnośląskiego Zagłębia Węglowego *Kwartalnik Geologiczny* **31/1** pp 57–68

[8] Kosewska M and Wróblewska K 1995 Chlor w gazie koksowniczym na poszczególnych etapach jego oczyszczania *Karbo-Energochemia-Ekologia* **40** pp 294–302

IOP Conf. Series: Earth and Environmental Science **174** (2018) 012012 doi:10.1088/1755-1315/174/1/012012

Perspective directions of the development of hydrotransport gravity mixers installations in the light of existing industrial solutions

Marcin Popczyk

Faculty of Mining and Geology, Silesian University of Technology, Akademicka 2, 44-100 Gliwice, Poland

marcin.popczyk@polsl.pl

Abstract. For fire prevention Polish hard coal mining has used mixtures of ash and water liquidshilts made on the basis of fine-grained energy wastes, which task is to fill free spaces by minimizing the possibility of oxygen access from mine air. The hydromixtures is prepared in surface mixer installations and gravitationally transported by a pipeline system to underground workings. Considering the availability of fine-grained energy waste in the immediate vicinity of hard coal mines, it should be assumed that this type of fire prevention will be successfully used in the future. The effectiveness of this method depends primarily on the efficiency and optimal operation of the mixing plant. The paper presents the evolution of mixing systems used in the last decades in hard coal mines with the division into generations, as well as the prospective directions of their development.

1. Introduction

Exploitation of hard coal in Poland is inseparable from the occurrence of many threats. One of them is the threat of endogenous fires arising in the caving goafs created after the exploitation of the longwall system. The reason for these fires are coal residues, which are often in contact with oxygen in the mine air and heat up and, after a while, they become a process of slow burning. One of the methods of limiting this type of threat is the use of fire prevention consisting in introducing into the cavity space hydraulic ash and water hydromixtures using gravitational hydraulic transport in pipelines. The flow of water jets in this type of installations is a self-steering process. This means that flow parameters such as performance or hydrotransport velocity result mainly from the spatial layout of the installation route, in particular from the ratio of its length to the difference in the height of the inlet and outlet from the installation [1, 2, 3]. In addition to the shape of the pipeline route, hydrotransport performance is also affected by the parameters of the water mixtures, such as the density and viscosity resulting from the proportion of water to fine-grained material (fly ash). If the pipeline route is shaped, as the ratio of its length to the height difference of the inlet and outlet increases, the flow parameters of the backfilling mixture decrease, which results from the drop in the flow velocity. To ensure stable flow conditions of the mixture, the installation should be characterized by the so-called correct hydraulic profile, i.e. in which the pressure height line does not intersect the hydraulic profile of the installation [2]. Otherwise, the flow of the mixture in the installation is disturbed and in the extreme case it is impossible. The standard parameter defining the consistency and the "transportability" of a liquid mixture is pourability [4]. On the basis of practical experiments and multiple measurements verified both in laboratory and industrial conditions, it can be stated that mixtures intended for caulking gullets (voids characterized by small dimensions) should have spreading in the range from 180 mm to 220 mm, and mixtures intended for filling voids of large dimensions, especially those having the character of a self-extinguishing filling, should have a smaller leveling in the range from 140 mm to 180mm [5].

Increasing the level of the mixture is achieved by increasing the proportion of water in the mixture. The effect of increasing the proportion of water in the mixture is to reduce the concentration of solids in the mixture, reduce the density of the mixture, lower the viscosity, and increase the setting time, deteriorate the strength properties of the solidified mixture, and in extreme cases even the loss of solidification. From the point of view of hydrotransport, the mixture with lower concentration of solids ensures lower resistance of flow in the installation, and thus provides higher flow velocity and a greater range of transport - which is important especially in the case of a significant horizontal distance of works from the filling shaft or a small level difference between the surface and the place of carrying out the filling works. In the caving goafs, the mixtures with the greater leveling provide greater penetration and better filling of the goaf, especially in the case of small porosity [2, 4].

On the other hand, the high leveling mixture contains a much larger amount of water than can be bound by fly ash contained therein, which drains into underground workings and additionally charges the mine drainage system. Therefore, the most important element in the correctness of carrying out the process of hydraulic fluid handling and transport is an exact mixer system that performs a mixture of water with specific rheological parameters [1].

2. Characteristics of the hydrotransport gravity installation

Currently, suitable gravity hydrotransport installations on hard coal mines in Poland are used to make hydraulic backfill or hydromixtures of ash and water. The hydraulic stowing is a mixture of water and sand with the possible addition of crushed gangue and its main purpose is to protect the surface from the effects of underground mining on the surface. The hydraulic stowing can also be used for making filling jacks and eliminating headings[2, 6]. Currently, due to the lack of longwall excavations using hydraulic backfilling, this direction of gravity installations should be considered historical and non-development and is not analyzed in the paper.

The second direction of the use of gravity installations is the production of ash and water mixtures in them, which are the material for such mining technologies: fire prevention, reconstitution of caving goafs, liquidation of excavations, making jacks, solidified fill. The basic materials used in such installations are fine-grained energy wastes, coal slimes and flotation wastes. The efficiency and correctness of the operation of the mixing and pipeline transport system depends on the following elements [4]:

- limited hydrotransport coverage (length of the route),
- the level of homogenity of the mixture,
- dosing accuracy of ingredients,
- minimization of dewatering,
- the possibility of dosing and mixing of different fine-fraction waste.

Most of these elements are associated with the mixing system used in the surface part of the installation, therefore the method of dosing and mixing individual components is the key from the point of view of the process correctness. In the last few decades, mixing installations producing fine-grained liquid water mixtures have evolved from the simplest, to a very small extent, controllable mixing systems, to the currently highly advanced dosing, monitoring and mixing systems realized with the use of full industrial automation. Table 1 shows the development of mixing plants for fine-particle materials with water over the last decades divided into generations.

Table 1. The evolution of mixing systems of fine-particle materials in recent decades in Polish hard coal mines.

Generation	Mixer type	Feeding to pipeline	Pipe diameters (mm)	Automation, sensorisation, dosage	Feeding with fine-grained material
I	Flow (tubular, counter current)	gravitationally through the wash hopper of the hydraulic filling plant	185	No full control	Tank trucks, rail tankers
II	flow (tubular, counter current)	Gravitational through: - the wash hopper of the hydraulic filling plant - to a pipeline led out to the surface	185	No full control	Tank trucks, rail tankers Storage silos
III	One- and two-shaft mixers	Pumping through an intermediate tank	Shaft 185 Workings 150	Manual water control, visual assessment of density	Tank trucks, railway tankers in cooperation with retention silos
IV	One- and two-shaft mixers	Pumping through an intermediate tank or directly	Shaft 185 150 Workings 150	Automatic dosing control based on the densometer	Retention silos
V	Duo installation. Combined fine-grained and sand mixing system	Variable layout: - feeding the hydraulic backfill by gravity - a fine-grained mixtures by pump	Shaft 185 150 Workings 150	Full measurement automation: - density - flow rate - pressure	Large storage retention silos, surface storage of sand

3. Mixing systems of fine-particle materials with water

Among the used systems of mixing dry fine particles with water can be distinguished:

- counter-current,
- flow,
- flow with the use of a rotor pump,
- special mixers using a feed pump.

3.1. Counter-current mixing system

One of the first systems of mixing fine-grained materials with water in order to obtain a hydromixture was the so-called mixing system. "Counter-current" or "direct" (figure 1). The fine-grained material (fly ash) was pressed directly in a pneumatic way using compressed air from cement cars or cement wagons to the inlet of the transport pipeline with simultaneous water supply. Pre-mixing occurred at the point where the dry material meets the water and then the mixing process proceeded during the flow in the transport pipeline. Such a mixing system was characterized by low emergency design and a simple operating principle. The basic disadvantage of this system was the lack of possibility to control the density and level of homogenisation, and the efficiency of work depended only on the experience of service staff. The guarantee of trouble-free operation of the installation was the administration of significant quantities of excess water to underground workings. An additional disadvantage was the feeding of significant amounts of air into the pipeline, which disturbed the laminar flow of the water mixtures, causing pulsation and hydraulic impacts during the flow in the pipeline.

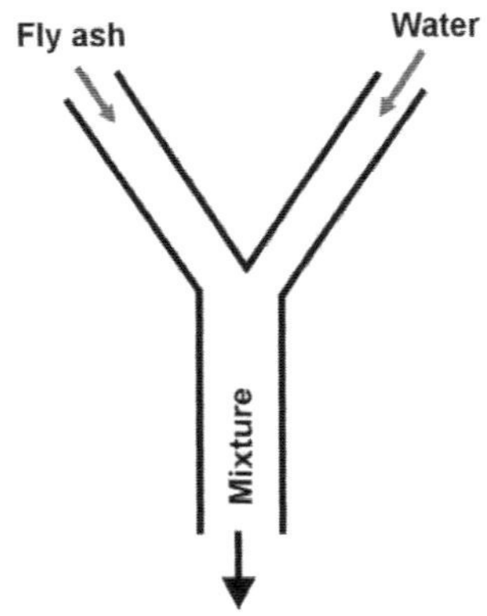

Figure 1. Counter-current mixing system.

3.2. Flow mixing system

Another type of mixing systems used is a flow mixer, whose cross-section and view was shown in figures 2, 3 and 4. The operating principle of the mixer consists in feeding the dry material (fly ash) pneumatically from cement cars or cement wagons directly to the mixer. In the mixer, the ash flow is through the perforated section of the pipe through which the water under pressure in the outer coat of this section gets. The mixing takes place on a longer road than in a counter-current mixer and the outgoing hydromixture has a homogeneous character immediately after leaving the mixer. The efficiency of the mixer depends on the pressure of the feed water and the amount of pneumatic delivery of dry fly ash.

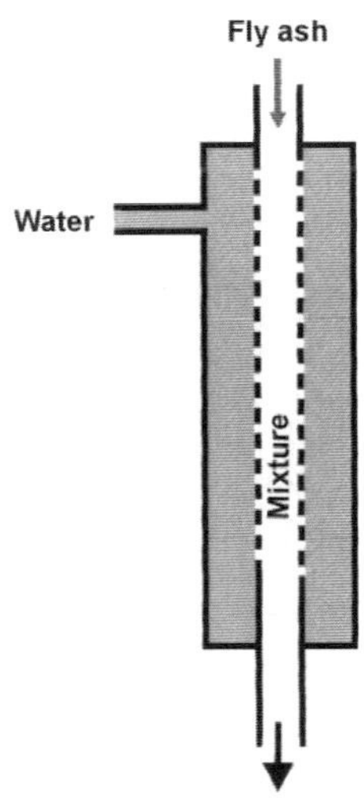

Figure 2. Flow mixer using a water coat.

Increasing the efficiency of the hydromixture feeding is realized by parallel installation of several sets of such a mixer as in figure 3. This solution allows simultaneous unloading of several cement cars or cement tanks.

Figure 3. Built-up set of four flow mixers.

The main advantage of this device is its mobility and simplicity of design, which allows its use also in field work. The example of this application can be liquidation by filling with fine-tuning hydromixtures surface sites as in figures 4 and 5.

Figure 4. Flow mixer built-up in the field.

Figure 5. Place of feeding of fine-grained mixture to the surface.

IOP Conf. Series: Earth and Environmental Science **174** (2018) 012012 doi:10.1088/1755-1315/174/1/012012

3.3. Flow mixing system using an impeller pump

In the case of using a mixer installation away from the place of feeding to the vertical part of the pipeline installed in the gravity transport shaft, it is necessary to mix and pump the liquid mixture. In this case, mixer systems equipped mostly with rotary pumps adapted for hydrotransport of granular materials of the PH or HC series are used (figure 6). This solution allows for a high degree of homogenization of the water mixtures, reduction of the amount of excess water, obtaining a high-density consistency and the use of components dosing control systems.

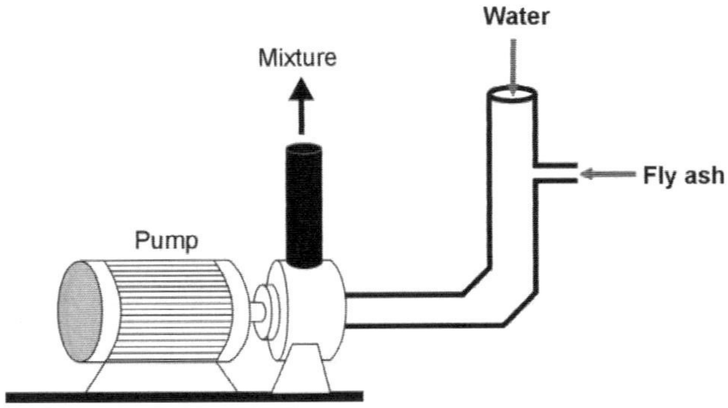

Figure 6. Mixer system using a rotary pump.

3.4. Special mixer system using a feed pump

The currently built installations for small-fraction hydromixtures are based on ensuring the appropriate quality of the produced water mixtures in terms of density, consistency and minimization of excess water. In addition, these installations are based on the addition of dry material to the mixer from retention tanks and not directly from car tanks or wagon cisterns.

Such systems ensure:

- accuracy of homogenisation of the mixture,
- dosing accuracy of ingredients,
- the possibility of using process automation,
- reduction of operating costs of the installation,
- rational water management,
- the ability to create mixtures for current technological needs,
- low investment costs for container installations,
- elimination of a significant amount of air from hydromixtures in comparison to flow installations.

The system operation process is divided into two stages: the mixing stage and the feeding stage for the hydrotransport pipeline. The mixing system is based on a high-efficiency mixer, usually with screw construction, working in a continuous cycle. The ready-mixed liquid mixture coming out of the mixer is usually stored in a storage reservoir before it is injected into the transport pipeline. The feeding system is equipped with a screw pump or a rotor pump that draws in a hydromixture from the retention tank and feeds it into the transport pipeline. The use of a retention tank allows to eliminate disturbances in the operation of the installation due to the different efficiency of the mixer and the pumping system. Examples of such an installation are shown in figures 7-9.

Figure 7. Retention tanks for fly ash.

Figure 8. Screw mixer.

Figure 9. Hydromixture's retention tank.

The use of mixer systems with the use of feed pumps allowed for full automation of the dosing process and preparation of fine-grained hydromixtures with the possibility of building such a plant in field conditions as shown in figures 10-13.

Figure 10. View of the local mixer installation.

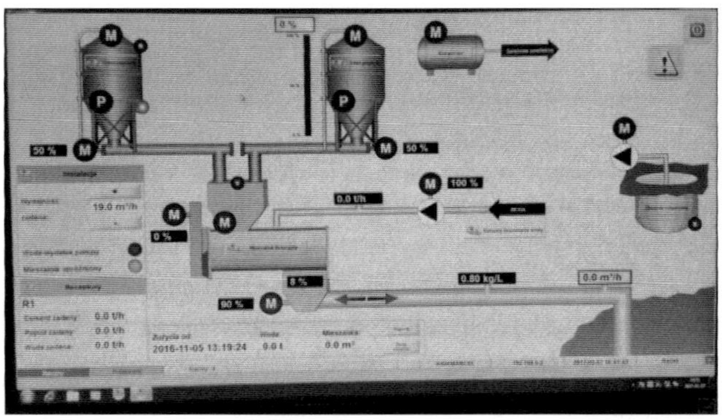

Figure 11. Installation control and monitoring module.

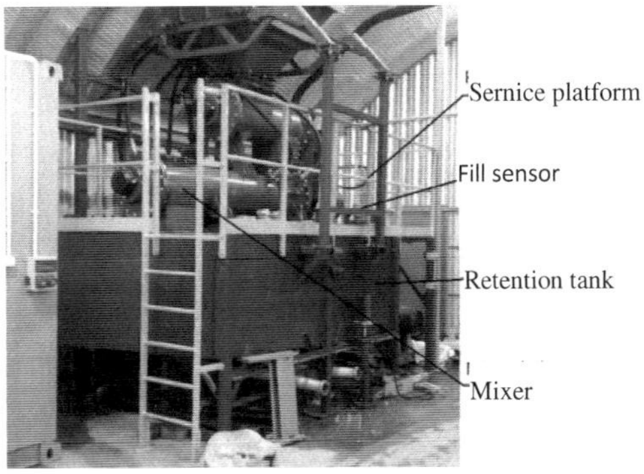

Figure 12. Elements of the mixer and feeding system.

Figure 13. The virtual model of cement tank unloading station.

4. Future installations for the preparation and transport of two-phase hydromixtures

As mentioned in the introduction of the paper, installations making hydraulic backfill made of water and sand as well as grained coarse material such as mine stone are now in liquidation or used to a negligible extent mainly for the liquidation of corridor workings. However, this can be implemented only in the mine areas with access to such installations. Considering the cost of building such an installation, one should not expect such an investment in the coming years. An alternative solution is to combine a mixer installation of fine-grained materials with a mixing system of water and grained coarse material to form a so-called mixer "Duo-installation". The first such concepts and projects have already been created and will be implemented in the coming years. Below, in figures14 and 15, a model of such an installation is shown which is divided into two modules.

The first duo-installation module that produces fine-grained liquidshare is equipped with the following basic elements:

- storage silos for ashes,
- ash transport system from silos,
- ash buffer tank,
- mixing system, along with the metering and ash feeding system,
- buffer tank for the mixture,
- mixture pump,
- hydraulic installation for feeding the mixture.

The second module of the duo-installation for the production of a water-sand mixture is equipped with the following basic elements:

- sand storage yard,
- sand bunker,
- sand screen,
- dispenser belt feeder,
- conveyor belt,
- sand washing station.

The principle of the functioning of such a duo-installation consists in the daily work of the module for preparing fine-grained hydromixtures for fire prevention, while the use of the hydraulic filling module will take place periodically, eg in the case of the need to quickly build a filling jack separating or insulating the fire region.

The capacity of the duo-installation can be estimated today for the production capacity of ash-water mixtures at the level of 120-200 m^3/h, and the production capacity of the water-clay mixture up to 300 m^3/h.

Figure 14. The view of the duo-installation model from the silo side of the fly ash.

Figure 15. The view of the duo-installation model from the side of the sand storage yard.

5. Summary

Mixing installations for the production of fine-grained hydromixtures have been successfully used in Poland for several decades, mainly for sealing the infestations of active walls in the framework of fire prevention. The produced hydromixture is transported by pipelines using gravity feed systems, and its flow parameters are mainly influenced by such factors as: the ratio of its length to the difference in inlet and outlet height from the installation and density and viscosity of the water mixtures due to the proportion of water to fine-grained material (fly ash). From the point of view of hydrotransport, the mixture with lower concentration of solids ensures lower resistance of flow in the installation, and thus provides higher flow velocity and a greater range of transport - which is important especially in the case of a significant horizontal distance of works from the filling shaft or a small level difference between the surface and the place of carrying out the filling works. On the other hand, the high leveling mixture contains a much larger amount of water than can be bound by fly ash contained therein, which drains into underground workings and additionally charges the mine drainage system. The first countercurrent and flow type mixing installations presented in the paper are characterized by the simplicity of operation and service, however, the hydraulic liquid from them contains a significant amount of excess water, which increases the costs of its service and significantly impairs the mine's drainage system.

Mixer systems introduced in recent years equipped with mixers supported by impeller pumps or feed pumps allowed for full automation of the mixing process as well as the dosing system of individual components ensuring the assumed parameters of the mixture in terms of consistency and density as well as flow parameters in the pipeline. The future of mixer systems is their version of the so-called "Duo" also includes the function of making a fine-grained mixture and a backfilling mixture made on the basis of sand and fine mine stone. An example of such a solution is presented in p. 4 of the paper. Such an installation can have a range of applications in various mining technologies, such as:

- caulking caving goafs,
- making jams and backfilling,
- liquidation of excavations with backfilling material and binding material,
- fast insulation of fire fields especially important in mines exploiting coal that is self-igniting.

The concept of duo-installation presented in the paper has the following advantages:

- accuracy of homogenisation of the mixture,
- dosing accuracy of ingredients,
- the possibility of using process automation,
- reduction of operating costs of the installation,
- rational water management,
- the ability to create mixtures for current technological needs,
- low investment costs for container installations,
- elimination of a significant amount of air from hydromixtures compared to flow installations,
- low investment costs of the module for making the water-sand mixture compared to the costs of constructing a classical backfilling plant.

References

[1] Popczyk M 2017 The impact of changes in the rheological parameters of fine-grained hydromixtures on the efficiency of a selected industrial gravitational hydraulic transport system. *IOP Conference Series: Materials Science and Engineering* **268** conference 1

[2] Palarski J 1997 Fill transportation and sedimentation mechanisms for stope filling *9th International Conference on Transport and Sedimentation of Solid Particles* Kraków

[3] Plewa F, Mysłek Z and Strozik G 1997 Parametry hydraulicznego doszczelniania zrobów zawałowych odpadami drobnofrakcyjnymi *III Szkoła Geomechaniki* Ustroń pp 2003–12

[4] Popczyk M 2002 *Ekologiczne i górnicze aspekty przemysłowego wykorzystania odpadów energetycznych i poflotacyjnych w podziemiach kopalń* Praca doktorska Wydział Górnictwa i Geologii Gliwice

[5] Plewa F and Mysłek Z 2001 *Zagospodarowanie odpadów przemysłowych w podziemnych technologiach górniczych* (Gliwice: Wydawnictwo Politechniki Śląskiej)

[6] Palarski J 1981 *Hydrotransport* (Warszawa: WNT)

[7] Różański Z, Suponik T, Matusiak P, Kowol D, Szpyrka J, Mazurek M and Wrona P 2016 Coal recovery from a coal waste dump. *E3S Web Conf.*, **8 01052** DOI: http://dx.doi.org/10.1051/e3sconf/20160801052

[8] Wrona P, Różański Z, Pach G, Suponik T and Popczyk M 2016 Closed coal mine shaft as a source of carbon dioxide emissions *Environmental Earth Sciences* **17(15)** DOI:10.1007/s12665-016-5977-7

IOP Conf. Series: Earth and Environmental Science **174** (2018) 012013 doi:10.1088/1755-1315/174/1/012013

Fire hazard in coal waste dumps – selected aspects of the environmental impact

Zenon Różański

Faculty of Mining and Geology, Silesian University of Technology, Akademicka 2, 44-100 Gliwice, Poland

zenon.rozanski@polsl.pl

Abstract. In areas of coal basins all over the world, wastes are produced along with the process of coal exploitation. Formerly, the wastes were deposited to waste dumps in large amounts. The presence of coal waste dumps at the surface is strictly related to their environmental impact. One of the more significant problems related to the environmental impact is constituted by spontaneous fires. This paper presents the selected aspects related to the negative impact of spontaneous fires on the environment. Based on own experience and scientific literature, the causes and the mechanism behind the formation of fires as well as the factors impacting the development of fires were presented. The paper also concerns the environmental impact resulting from the process of leaching of contaminants from the wastes to surface and ground water as well as fire-related dust and gas emissions to atmosphere. It has been found that the emission of substances to the atmosphere is the most disturbing to the environment. The emissions include significant amounts of toxic substances and greenhouse gases.

1. Introduction

Underground exploitation of mineral resources and the production of significant amounts of waste rock that are extracted to the surface are inseparable. In hard coal mining, the wastes are constituted by barren rock accompanying coal seams. The wastes which are produced alongside the exploitation and processing of coal are commonly called mining wastes, although the name encompasses also some power production wastes, that is, wastes produced in the process of coal combustion. These wastes include i.a. ashes, slags and fly ash.

Due to the high output of numerous mines in coal basins, enormous amounts of mining wastes were produced, and the problem of neutralising these wastes emerged. It was only at the end of the previous century that large-scale activities aimed at the use of the mining wastes were taken up. Beforehand, a larger part of the wastes was deposited at the surface. Numerous dumps were established by the coal mines or in the nearest vicinity of the mines, where enormous amounts of barren rock and often other kinds of waste materials were deposited.

There are hundreds of objects constituting the location of coal mining waste in the coal basins around the world [1, 2]. There are approx. 250 such objects in Poland, yet there are countries in which this number is even higher. For example, in China, the number of dumping grounds on which gangue from bituminous coal mines was deposited exceeds 1700, and their presence constitutes a significant environmental problem [3].

Besides the mining dumps' disadvantageous impact on the landscape, the dumps may have a negative impact on the environment. Their chemical composition and the possibility of leaching of substances are a source of pollution of surface water, ground water and soil [4, 5, 6, 7]. Certain amounts of fine-grained material contained therein may be easily transported by wind thus contributing to the increased dust levels in the atmosphere. Their most disturbing effect is, however, exhibited upon the emergence of fire in these objects. Fires of mining waste dumps are a typical and common phenomenon. Such fires constitute a source of emissions of gas substances to the atmosphere and increase the dust levels in air. Due to the high temperatures accompanying the fires, they contribute to the general safety hazard posed to people present within the outline of the dumps [8, 9, 10, 11, 12, 13, 14, 15]. The main cause of the fires is the spontaneous process of low-temperature oxidation of coal and pyrite contained in the wastes from Carboniferous deposits. This process is accompanied by an increased production of heat and increased temperatures. Most of the old cone-shaped dumps, where the wastes were built by free deposition, have formerly been subject to spontaneous combustion. In some of the dumps, the thermal activity continues since several dozen years until now.

The article presents selected environmental aspects related to the presence of mining waste dumps and the occurring fires thereof. Own research results and the results obtained by other researchers were used in the paper.

2. Causes of fires of coal waste dumps

The waste material deposited in waste dumps consists mainly of gangue. However, it also contains substantial amounts of combustible constituents. Along with gangue, coal is deposited in waste dumps in the form of bands or as a constituent of rocks that coexist with coal seams, such as coal clay or shale. A significant constituent of coal waste is also pyrite [16]. Both coal and pyrite are chemically active in relation to oxygen, and the reactions they go into are exothermic. Moreover, other combustible materials may also be deposited together with gangue in waste dumps: fragments of conveyor belts, wooden components of mine supports, etc.

The presence of significant amounts of combustible materials in coal mining waste dumps is the main cause of the occurrence of fires at these sites. Depending on the cause of fire occurring in waste coal deposition sites, fires may be classified as:

- exogenous,
- endogenous.

Exogenous fires arise as a result of external heat coming from sources such as bonfires, grass burning, welding operations, hot ash disposed on the dump with other mining waste, etc. Such fires may be prevented relatively easy by avoiding the activities mentioned above. Lack of awareness makes the common practice of grass burning a frequent cause of waste dump fires.

Endogenous fires are the result of natural low-temperature process of the oxidation of a carbonaceous or other substance, e.g. pyrite. Oxygen access to the bulk of the dump promotes oxidation of carbonaceous substances, which is accompanied by heat generation. The mechanism of initiating endogenous fires is similar to that of an underground fire [17]. The conditions that must occur simultaneously for a fire to start include:

- appropriate amount of substance chemically active in contact with oxygen in air,
- access of air to the bulk of the dump,
- heat accumulation inside the dump (amount of heat generated in exothermic reactions exceeds the amount of heat given off to the environment).

2.1. Mechanism of initiation of endogenous fires

Despite many years of research, the process of coal self-heating, which is the main cause of coal waste dump fires, has still not been fully comprehended. According to Banerjee and Pone, oxidation of carbonaceous substances proceeds in four stages [18, 19]:

1) physical adsorption of oxygen, initially accompanied by a rise in temperature (by 30 to 50°C);

2) chemical absorption (temperature above 50°C), which leads to the formation of unstable, oxidized hydrocarbon compounds and the so-called peroxygen complexes,

3) decomposition of oxidized hydrocarbons and peroxygen complexes to provide additional oxygen amounts at subsequent stages of oxidation (temperature range above 70°C),

4) process acceleration, temperature increase to ignition point above 150°C (ignition point depends on coal rank).

These processes may lead to temperature increase to even above 1000°C. In the meanwhile oxygen consumption is accelerated and gaseous products are released violently.

The course of self-heating and spontaneous ignition of coal is shown in figure 1. In the first stage of fire development, self-heating of coal (waste material) proceeds and the temperature slowly rises. This period is called the induction (or incubation) period. When temperature reaches critical value (T_{cr} = ca. 60-80°C), the rate of reaction increases rapidly. When the conditions listed above are met, temperature increases to ignition temperature T_i (the value of which depends on the type of coal) at which ignition occurs and in consequence fire starts. If critical temperature is not reached or if conditions change to promote heat abstraction, then a cooling period starts. The self-heated coal undergoes weathering.

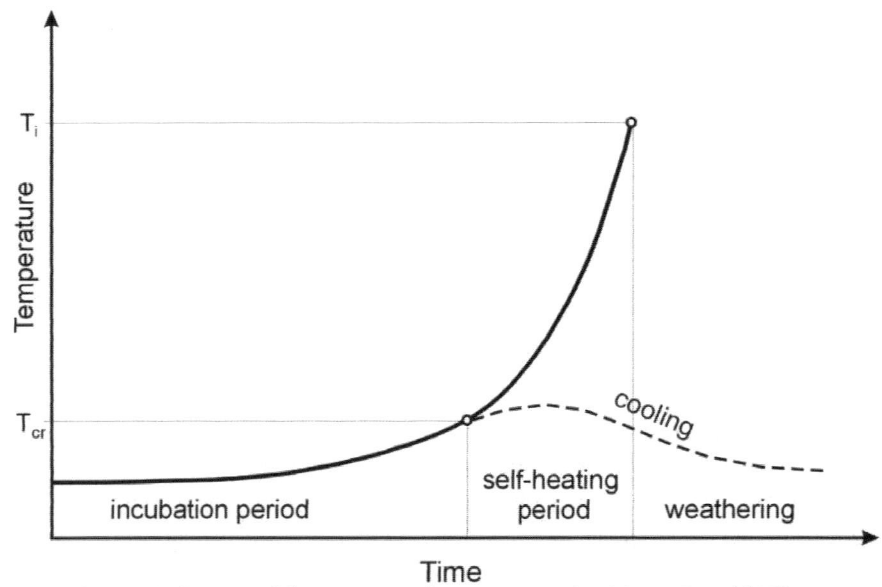

Figure 1. Course of the process of spontaneous ignition of coal [17].

Thermal phenomena result from many reactions of the constituents of the waste material or of secondary products of reactions that take place inside the coal waste dumps. These reactions include exothermic reactions that cause temperature increase, and endothermic reactions which, however, result in products that participate in subsequent exothermic reactions. The most important of them are listed in Table 1 along with enthalpy differences ΔH that are a measure of emitted (−) or absorbed (+) heat.

Due to the latent nature of the initial phase of endogenous fires, detecting them is difficult, and consequently combating them is more complex.

Table 1. Basic chemical reactions occurring during the self-heating of coal waste.

Chemical reaction	Enthalpy difference $\Delta H(kJmol^{-1})$
$C + H_2O = H_2 + CO$	+118.5
$CO + H_2O \rightarrow H_2 + CO_2$	–42.3
$CO + 3H_2 \rightarrow CH_4 + H_2O$	–206.0
$C + 2H_2 \rightarrow CH_4$	–87.5
$2C + O_2 \rightarrow 2CO$	–123.1
$C + O_2 \rightarrow CO_2$	–406.0
$C + CO_2 \rightarrow 2CO$	+159.9
$2FeS_2 + 7O_2 + 2H_2O \rightarrow 2FeSO_4 + 2H_2SO_4$	–2544.0

2.2. Factors influencing the fire hazard
There is a number of factors that have an effect on the probability of occurrence and intensity of thermal phenomena associated with self-heating. These can be classified into two groups:

- internal factors - which determine the reactivity of the waste material in relation to oxygen of air at low temperatures,
- external factors - related to ambient conditions or to features of the waste dump that have an effect on filtration capacity and the heat balance.

Reactivity of waste material depends on its physicochemical properties. The most important of these is material composition, which determines the content of combustible materials in the waste, in this case mainly coal and pyrite.

Grain size distribution in the material is also important. It depends on the origin of the waste material. Waste originating directly from mining or sorting operations contain coarse grains, or even large chunks. Mechanical processing and coal cleaning processes yield fine-grained material. Disposal of coal waste directly onto dumps by belt conveyors and rail cars, without any fire-preventive processing, which was a general practice in the past, caused the so-called grain segregation, which promoted self-heating [20]. When waste is disposed on the dump, fine grains tend to accumulate in the top and middle parts of the dump, while the coarse grains slide down the slope and gather in the bottom part of the dump or around it. The thus formed structure of the dump (figure 2) creates advantageous conditions for air penetration into the bulk of the dump, and consequently for initiating oxidation of carbonaceous materials. Fine waste from processing plants may cause partial self-sealing of the dump. This may help reduce air penetration into the dump, but on the other hand, processing waste contains more combustible constituents.

Susceptibility to spontaneous ignition of coal contained in waste also depends on carbon content (rank of coal). According to W. Gabzdyl [21] the study of the natural susceptibility of coal to spontaneous ignition requires, in the first place, the analysis of macerals. Vitrinite is the most susceptible to spontaneous ignition, whereas in the Upper Silesian Coal Basin the susceptible low-rank coals contain more than 30% of fusinite and semifusinite, that is macerals of the inertinite group. An important feature (in addition to petrographic composition) is carbon content (rank of coal) and macrostructure of the seam from which coal is derived.

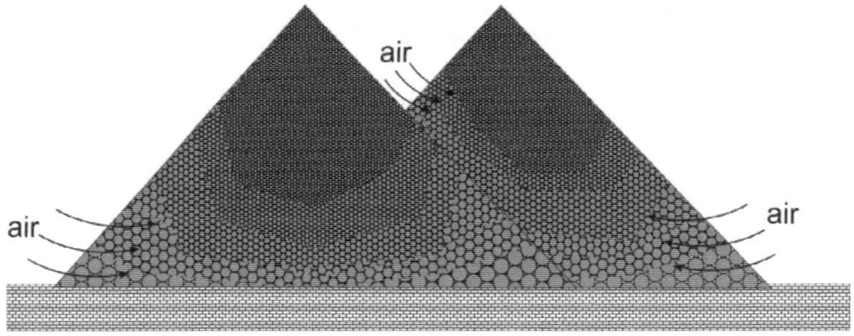

Figure 2. Structure of cone-shaped dumps that promotes air penetration.

Low-rank coals containing thick-bedded vitrite are considered to be the most susceptible to spontaneous ignition. Susceptibility to spontaneous ignition depends on the content of mineral matter, particularly of sulphides (pyrite).

Higher rank coals, despite having higher carbon content, are characterised by lower susceptibility to spontaneous ignition. This is due to lower specific surface area and low volume of micropores [22], which reduces surface adsorption of oxygen. Coal rank also determines the composition of volatile components of coal, which has an effect on the ignition temperature of coal. Higher rank coals contain more highly aromatic hydrocarbons (with higher ignition temperature), and the volatile matter of lower rank coals contain more aliphatic compounds of lower ignition temperature. For the reasons listed above, spontaneous ignition is rare in coal waste dumps of coking coal mines in the Jastrzębie region.

Important external factors that affect the susceptibility to spontaneous ignition of waste dumps include the applied haulage, dumping and fire prevention methods. They have an effect on the air tightness of the waste dumps. As mentioned before, dumps formed in an uncontrolled manner have an unfavorable structure as regards fire hazard. The applied methods of dumping and fire prevention must ensure the highest possible compaction of the deposited waste material. It is assumed that the minimum value of the compaction factor when banks of coal waste are formed is $I_s = 0.95$ [23]. Waste dumps formed nowadays are much less susceptible to the occurrence of thermal phenomena. Dumping of waste material now does not consist in uncontrolled deposition, but rather in planned construction of dumps with elimination, or at least minimization, of factors that promote self-heating and spontaneous ignition. Factors that affect the filtration capacity of a waste dump, which also result from the manner of dump formation, include its shape and height. This is related to the action of wind on the slopes of the dump. The most exposed to fire hazard are slopes of the dump that are high, steep and situated on the windward side. For this reason, at particular risk are the slopes of conical dumps of significant height and of high angle of slope, due to both wind action and grain segregation described before. In Upper Silesia (particularly in the Rybnik Coal Area) there are many examples of conical waste dumps that have already burned out or that have been thermally active for several dozen years (figure3).

Waste tips that are level with the ground, and therefore have no slopes, are much less susceptible to self-heating because of the reduced surface exposed to air penetration induced by winds.

Self-heating is also affected by other weather factors. Solar radiation causes heating of the uppermost layers of waste material, which affects the heat balance of the waste dump. Rains, particularly heavy rainfalls, cause damage to the slopes by forming fissures, furrows or breaches which facilitate air access to the bulk of the dump [20]. Atmospheric pressure variations (decrease, increase) cause a phenomenon that is called baric aeration [24]. Increased pressure forces air into the bulk of the dump. Under lower pressure air moves in the opposite direction.

Figure 3. Examples of cone-shaped waste dumps in Poland – local absence of snow cover is an evidence of thermal activity of waste material.

The scope and manner of applying fire prevention is a very important factor influencing the susceptibility of waste dumps to spontaneous ignition. At present, waste dumps are constructed in a planned manner and methods and means are applied to effectively reduce the risk of spontaneous ignition. Methods of prevention depend on the conditions determining the probability of fire occurrence mentioned before. They are based mainly on reducing the concentration of combustible matter in the deposited material by adding inert material, on reducing the capacity to accumulate heat and on cutting off oxygen access to the bulk of the dump.

The evaluation of the simultaneous effect of the factors mentioned, consisting in assigning numerical weights to these factors, may form a basis for assessing fire hazard in coal waste dumps [20, 23]. A modified method of assessing fire hazard developed by the Central Mining Institute (GIG) takes into account the following factors [23]:

- type of dump,
- susceptibility of waste to self-ignition,
- scope and frequency of monitoring,
- compaction factor,
- physical properties of waste (compactibility).

At present, of the 250 or so objects formed of coal waste in Polish coalfields, at least a dozen or more manifest thermal activity. This results in adverse environmental effects.

3. Environmental impact of burning coal waste dumps

The existence of coal waste dumps is in itself a source of environmental impact. This is manifested by affecting the soil and water environment and the air surrounding the objects.

Waste deposited in dumps is a source of contaminants that are washed out into surface and ground waters [25, 26, 27].

In the case of managing coal waste on ground surface (which includes waste dumps), the results of the analysis of water extracts thereof are compared with the permissible values specified in the Ordinance of the Minister of Environment of 24 July 2006 on the conditions of discharging wastewater to waters and soil and on substances particularly harmful to aqueous environment [28].

Analysis of water extracts of coal waste in the Upper Silesian Coal Basin [27] has shown that the determined basic ions (sodium, potassium, chlorides, sulphates) are washed out in quantities much lower than those permissible for wastewater discharged to soil or waters and specified in the Ordinance of the Minister of Environment (table 2). Some mining waste, however, when used in the environment, due to chlorides content, may have an adverse effect on vegetation growth and may lead to contamination of surface waters, or even of ground waters.

Table 2. Mean and maximum results of the determination of pH and composition of water extracts of waste from mines of the various regions of the Upper Silesian Coal Basin [27] and permissible values [28]

Region	Concentration	pH	Na^+	K^+	Cl^-	SO_4^{2-}
			mg·dm^{-3}			
Central	mean	8.27	56.0	2.8	62.3	9.8
	maximum	9.15	161.0	5.5	255.9	32.5
East	medium	8.11	99.6	4.9	134.7	25.5
	maximum	9.24	435.0	13.4	867.0	52.7
Bytom	mean	8.28	38.0	3.8	24.9	6.3
	maximum	8.92	63.0	6.1	72.2	25.9
Rybnik-Wodzisław	mean	8.74	69.0	2.4	68.6	11.2
	maximum	9.06	102.0	4.2	110.2	14.8
Total	medium	8.28	66.6	3.5	77.1	13.7
	maximum	9.24	435.0	13.4	867.0	52.7
	permissible	6.5-9.0	800	80	1000	500

Leachates that form in mining waste deposits constitute a real threat for the soil environment and for ground waters because of high loads of heavy metals (table 3).

The processes of leaching water soluble components proceed at varying rates that depend on physicochemical transformations that occur in Carboniferous rocks. Physical, chemical and biochemical weathering of waste gathered in the dumps is of high importance in this process. The content of components in freshly deposited waste is relatively low. Much higher quantities of contaminants arise in the physicochemical process of waste weathering [29].

Thermal processes occurring during self-heating and fires of waste dumps are also significant. The degree of thermal conversion of waste has an effect on metal leachability. As shown by the example of waste dump no. 4 at the Nowy Wirek Coal Mine, much more elements (Al, Co, Cr, Cu, Ni) are leached from samples of deeply thermally transformed samples (sinters) than from samples of thermally intact (unburned) waste and moderately thermally transformed (burned out) waste. This is due to the presence of significant amounts of glaze in sinters, such glaze being subject to devitrification and release of metals contained therein [30].

Table 3 shows results of determination of trace elements in samples of soil taken in the vicinity of a waste dump where the deposited material had undergone deep transformation as a result of thermal processes associated with fires. The table also shows similar results for water samples taken from a lagoon formed after excavation of part of the waste material. For most of the listed elements, the content thereof in water from the lagoon is several-fold higher than the values reported for surface waters in the region of Ruda Śląska [31]. As concerns the soil studied, limits (specified in the Ordinance of the Minister of Environment of 9 September 2002 on soil and ground quality standards) have not been exceeded, with the exception of Ba and Zn. However, the values determined were much higher than those of the background specified in the Geochemical Atlas of the Upper Silesia for the following elements: Ba, Co, Cr, Cu and Ni.

Table 3. Content of trace elements in soil and water samples taken in the vicinity of the waste dump in Ruda Śląska [32].

Content of the elements	Al	As	Ba	Cd	Co	Cr	Cu	Fe	Ni	Pb	Zn
in soil	nd	<2 -23	282 -1120	<0.5-15.7	5-39	50-157	5-96	nd	14-102	8-298	70-1930
in water	0.045	0.559	0.423	-	0.031	0.057	0.157	0.294	0.217	nd	0.921

nd - no data(analysis not performed)

Another environmental component that is exposed to the impact of coal waste deposits is air. Adverse effects are related to emissions of dust, greenhouse and toxic gases [8].

Waste dumps contain fine-grained material which is carried away by wind and causes particulate pollution in air. This can easily be abated by biologically reclaiming the waste dumps. However, there are still many objects that have not been reclaimed at all or that have been reclaimed to an insufficient extent only.

If fires occur in waste dumps, pollutant emissions to air increase substantially. Dust content in air may attain grievous levels. Because of increased temperature and intensified evaporation, the material in the waste dump is freed of moisture. Dry particulates levitate in air more easily. If in the course of a fire, coal contained in waste is transformed into ash, the tendency to be carried away by wind is intensified, particularly during operations associated with combating fire, removal and transportation of burned material or reclamation works (figure 4).

Dust from mining waste dumps does not only affect human health, but can also contaminate the adjacent soil. Bianhas studied such impact of waste dumps on the soil. An increased content of Cu, Pb, Cd, Cr and Zn was exhibited in the soil in vicinity of a mining waste dump that was on fire, even at a distance of 100 m from its slopes [3].

Figure 4. Dusting in waste dump during fire combat and other operations.

Gases evolving during a fire are emitted from the surface of the thermally active waste dump into the atmosphere (figure5). The number and amount of emitted pollutants depend on many factors, among them:

- chemical composition of the waste,
- oxygen and water access to the bulk of the dump,
- stage of fire development and depth,
- temperature of waste material,
- gas transport in the interior of the dumping ground,
- weather factors, e.g. atmospheric pressure, wind etc.

The main gaseous substances emitted during endogenous fires are the products of reactions listed in table 1. These include: CO_2, CO, water vapour, CH_4, H_2. However, there are many other products resulting from other heat-induced reactions. As the waste material contains sulphur, sulphur dioxide SO_2 and hydrogen sulphide H_2S are emitted. Evidence of that is yellow efflorescence on the surface of the waste dump (figure 6).

Figure 5. Gases emission from the surface of burning waste dumps.

Figure 6. Sulphur efflorescence in gases outflow places.

Example results of chromatographic determination of main gas concentrations in samples taken from the fire-infested area of the Wrzosy dump of Rydułtowy-Anna Coal Mine are shown in table 4. Concentration of toxic carbon monoxide above the surface of the waste dump within its thermally active area was above the permissible limits.

Table 4. Concentrations of gases in the thermally active area measured at the outlets of boreholes and 0.5 m above the surface of the Wrzosy waste dump of the Rydułtowy-Anna Coal Mine.

Measurement location	Average fire gas concentrations, (vol.%)								Interior temperature (°C) (depth)
	O_2	CO_2	CO	H_2	CH_4	C_2H_6	N_2	Σ	
Borehole 1	12.45	6.15	4.4	1.185	0.728	0.095	74.2	99.21	420 (1.0m)
Borehole 2	19.2	0.975	0.2805	0.0515	0.0515	0.01	78.95	99.52	260 (0.5m)
Borehole 3	12.9	5.01	1.1	0.226	0.226	0.07	79.9	99.43	358 (0.5m)
Borehole 4	13.6	6.345	0.1495	0.1665	0.1665	0.04	78.9	99.37	268 (0.3m)
Borehole 5	14.3	5.53	0.5	0.277	0.277	0.06	78.55	99.49	370 (0.2m)
Borehole 6	20.4	0.62	0.005	0.028	0.028	0	78.9	99.98	125 (0.2m)
0.5m *	20.8	0.11	0.005	0	0	0	78.9	99.82	-
0.5m *	20.8	0.1	0.006	0	0	0	78.9	99.81	-
0.5m *	20.8	0.08	0.005	0	0	0	78.5	99.39	-
0.5m *	20.9	0.07	0	0	0	0	78.9	99.87	-

* above surface of the dump

Investigations [9, 11] have shown that fire-infested waste dumps are a major source of fugitive emissions of greenhouse gases. Potential rate of these emissions has been determined on the basis of investigations of the amounts of gases emitted from a defined surface area and concentrations of CO_2 and CH_4 in the gas mixture formed. The obtained measurement results indicate that from 1 m^2 of a highly fire-infested waste dump (interior temperature more than 400°C), under the assumption that the intensity of thermal phenomena is constant, ca. 8 $MgCO_2$ and 0.2 $MgCH_4$ is emitted to the air every year [9]. When the global warming potential of methane (GWP100 = 23 according to IPCC) is taken into account, this corresponds to equivalent carbon dioxide of 12.6 Mg CO_2e.

Limited oxygen availability, presence of water in pores and elevated temperature make the processes that proceed inside the waste dump similar to those of low temperature carbonization (at lower temperature) or to coking (at higher temperature) of coal. This is accompanied by evolution of subsequent gaseous substances. Under such conditions, compounds having genotoxic, mutagenic and carcinogenic properties are formed. These include polycyclic aromatic hydrocarbons (PAHs). This was confirmed by research conducted by P. Kuna-Gwoździewicz (GIG) [33]. Fire gases from one of waste dumps have been shown to contain two- and three-ring hydrocarbons, such as: naphthalene (Nap), acenaphthene (AcP), fluorene (Flu), phenanthrene (PA) and anthracene (Ant). The results of measurements are shown in table 5.

Table 5. Results of determination of PAH concentration in the area of thermal activity of waste dump [33].

PAH	Concentration($mg \cdot m^{-3}$)			RCC*
	average	minimum	maximum	
Nap	9.911	3.373	24.411	0.001
AcP	0.084	0.017	0.400	0.001
Flu	0.123	0.054	0.351	0.001
PA	0.129	0.054	0.314	0.001
Ant	0.004	0.001	0.025	0.010
Σ PAH	10.251	3.499	25.501	-

*relative carcinogenic coefficient (in relation to benzo[a]pyrene) according to [34].

The emissions of the PAHs referred to above from thermally active dumps were also confirmed by research conducted by Howaniec [35]. Insignificant amounts of other carcinogenic substances such as Fluoranthene (Flu), Pyrene (Pyr), Chrysene (Chr) have also been found in the fire-related gas. Their average concentrations, however, did not exceed 1 $\mu g \cdot m^{-3}$.

4. Conclusions

Coal mining waste dumps affected by fires constitute a significant environmental problem in coal basins all over the world. The prevention of spontaneous fires is still relatively difficult and costly, especially in case of old dumps with large volumes and disadvantageous shapes. Due to the above, it is difficult to eliminate this disadvantageous phenomenon.

Dumps affected by fires exhibit an adverse impact on the nearest environment. It manifests itself by leaching of pollutants, which are transported to the surface and ground water as well as the soil. Usually, however, the allowable concentrations specified in legal regulations are not exceeded. The largest problem, however, is constituted by dust and fire gases which are emitted to the atmosphere. The emission of fire gases is highly burdensome to the citizens of areas directly neighbouring thermally active mining waste dumps. The concentration of CO in the areas of thermal activity may reach values that are hazardous to humans. Moreover, an intense odour nuisance is present in the area of the dump. Even if not exceeding the maximal allowable concentrations in the atmosphere, the sulphur compounds and hydrocarbons (PAH) are sensible even at very low concentrations in air, due to the very low odour threshold. Significant amounts of greenhouse gases are also emitted to the atmosphere during fires.

To eliminate the hazard posed to the environment by the dumps affected by fire, it is necessary to apply fire prevention that should consist in the monitoring of these objects. This would allow for an early detection of thermal phenomena and the commencement of preventive actions that would reduce the factors contributing to the further development of fire (e.g. the removal of the combustible material, cutting off the air supply or cooling the heated zones).

References

[1] Gawor Ł 2014 Coal mining waste dumps as secondary deposits–examples from the Upper Silesian Coal Basin and the Lublin Coal Basin *Geology Geophysics and Environment* **40(3)** pp 285–9

[2] Zástěrová P, Marschalko M, Niemiec D, Durďák J, Bulko R and Vlček J 2015 Analysis of possibilities of reclamation waste dumps after coal mining*Procedia Earth Planet. Sci.* **15** pp 656–62

[3] Bian Z, Dong J, Lei S, Leng H, Mu S and Wang H 2009 The impact of disposal and treatment of coal mining wastes on environment and farmland. *Environ. Geol.* **58(3)** pp 625–34

[4] Skarżyńska K M 1995 Reuse of coal mining wastes in civil engineering - part 1: Properties of minestone. *Waste Management* **15(1)** pp 3–42

[5] Szczepańska J and Twardowska I 1999 Distribution and environmental impact of coal-mining wastes in Upper Silesia, Poland *Environ. Geol.* **38(3)** pp 249–58

[6] Suponik T and Blank 2014 Removal of heavy metals from groundwater affected by acid mine drainage *Physicochem. Probl. Miner. Process.* **50(1)** pp 359–72

[7] Pierzyna P, Popczyk M and Suponik T 2017 Testing the possibility of leaching salt debris obtained from underground excavations *E3S Web of Conferences* **18** EDP Sciences01033

[8] Sułkowski J, Drenda J, Różański Z and Wrona P 2008 Noticed in mining areas, environmental hazard connected with outflow of gases from abandoned mines and with spontaneous ignition of coal waste dump *Gospodarka Surowcami Mineralnymi* **24(3/1)**

[9] Różański Z 2009 The potential gas emission from a thermally active coal waste dump *Górnictwo i Geologia* **4(3)** pp 103–9

[10] Różański Z and Wrona P 2010 The fires of mine waste dumps and connected hazard for the environment, Polish experiences *Proceedings of ICCFR2 Second International Conference on Coal Fire Research* 19-21 May 2010 Berlin pp 52–8

[11] Łączny M J and Ryszko A (Eds.) 2012 *Praca zbiorowa w ramach projektu COOL's: System zarządzania likwidacją emisji CO_2 ze zwałowisk odpadów powęglowych* Central Mining Institute&Silesian University of Technology Katowice-Gliwice

[12] Fabiańska M J, Ciesielczuk J, Kruszewski Ł, Misz-Kennan M, Blake D R, Stracher G, and Moszumańska I 2013 Gaseous compounds and efflorescences generated in self heating coal waste dumps - a case study from the Upper and Lower Silesian Coal Basins (Poland). *Inter. Journ. of Coal Geol.* **116** pp 247–61

[13] Stracher G B, Prakash A and Sokol E V (Eds.) 2014 *Coal and Peat Fires: A Global Perspective: Volume 3: Case Studies–Coal Fires* Elsevier

[14] Smoliński A, Drobek L, Dombek V, Bak A 2016 Modeling of experimental data on trace elements and organic compounds content in industrial waste dumps *Chemosphere* **162** pp 189–98

[15] Surovka D, Pertile E, Dombek V, Vastyl M and Leher V, 2017 Monitoring of Thermal and Gas Activities in Mining Dump Hedvika, Czech Republic *IOP Conference Series: Earth and Environmental Science* **92(1)** p. 012060 IOP Publishing

[16] Skarżyńska K M 1995 Reuse of coal mining wastes in civil engineering - part 1: Properties of minestone. *Waste Management* **15(1)** pp 3–42

[17] Strumiński A 1996 *Zwalczanie pożarów w kopalniach głębinowych* (Katowice: Wydawnictwo Śląsk)

[18] Banerjee S C 1985 *Spontaneous Combustion of Coal and Mine Fires* (Rotterdam: A.A. Balkema)
[19] Pone J D N, Hein K A A, Stracher G B, Annegarn H J, Finkelman R B, Blake D R, McCormack J K and Schroeder P 2007 The spontaneous combustion of coal and its by-products in the Witbank and Sasolburg coaldfields of South Africa *Int. J. Coal Geol.* **72(2)** pp 124–40
[20] Urbański H 1983 *Rekultywacja techniczna usypisk odpadów kopalnianych ze szczególnym uwzględnieniem zwalczania pożarów.*Training proceedings SITG Katowice Poland
[21] Gabzdyl W 1984 *Petrografia węgla* (Gliwice: Wydawnictwo Politechniki Śląskiej)
[22] Cygankiewicz J 2001. Wpływ zawartości wilgoci w węglu na przebieg samozagrzewania *Górnictwo i Geologia* **6(3)** pp 37–47
[23] Gogola K, Bajerski A and Smoliński A 2012 Modyfikacja metody oceny zagrożenia pożarowego na terenach lokowania odpadów powęglowych *Prace Naukowe GIG Górnictwo i Środowisko* **2** pp 13–32
[24] Korski J, Henslok P and Friede R 2005 *Uwagi o przyczynach powstawania pożarów składowisk odpadów górniczych, zwalczaniu pożarów i profilaktyce przeciwpożarowe.* Seminary of The Strata Mechanics Research Institute of the Polish Academy of Sciences Kraków
[25] Szczepańska J 1987 *Zwałowiska odpadów węgla kamiennego jako ogniska zanieczyszczeń środowiska wodnego* (Kraków: AGH)
[26] Korban Z 2011 Problem odpadów wydobywczych i oddziaływania ich na środowisko, na przykładzie zwałowiska nr 5A/W-1 KWK „X" *Górnictwo i Geologia* **6(1)** pp 109–20
[27] Bojarska K and Bzowski Z 2012 Wyniki badań wyciągów wodnych odpadów wydobywczych z kopalń węgla Górnośląskiego Zagłębia Węglowego w aspekcie wpływu na środowisko *Górnictwo i Geologia* **7(2)** pp 101–13
[28] *Rozporządzenie Ministra Środowiska z dnia 24 lipca 2006 r., w sprawie warunków, jakie należy spełnić przy wprowadzaniu ścieków do wód lub do ziemi, oraz w sprawie substancji szczególnie szkodliwych dla środowiska wodnego* (Dz.U. Nr 137, poz. 984)
[29] Twardowska I, Szczepańska J and Witczak S 1988 *Wpływ odpadów górnictwa węgla kamiennego na środowisko wodne. Ocena zagrożenia, prognozowanie, zapobieganie* (Wrocław: PAN)
[30] Nowak J 2011 Influence of thermal changes degree of the coal wastes on leaching substances into environment *Górnictwo i Geologia* **6(4)** pp 59–70
[31] Atlas Geochemiczny Górnego Śląska 1995 (Warszawa: PIG)
[32] Hanak B and Nowak J 2010 Estimation of chosen metals concentration and components soluble in soils and water found of KWK Nowy Wirek *Górnictwo i Geologia* **5(4)** pp 115–23
[33] Kuna-Gwoździewicz P 2013 Wielopierścieniowe węglowodory aromatyczne w gazach stref ekshalacyjnych wybranego termicznie aktywnego składowiska odpadów wydobywczych z górnictwa węgla kamiennego *Journal of Sustainable Mining* **12(1)** pp 7–12
[34] Nisbet I C T and LaGoy PK 1992 Toxic (TEFs) for polycyclic aromatic hydrocarbons (PAHs) *Reg. Toxicol. Pharmacol.* **16(3)** pp 290–300
[35] Howaniec N, Kuna-Gwoździewicz P and Smoliński A 2018 Assessment of Emission of Selected Gaseous Components from Coal Processing Waste Storage Site *Sustainability* **10(3)**

IOP Conf. Series: Earth and Environmental Science **174** (2018) 012014 doi:10.1088/1755-1315/174/1/012014

Design of underground mine voids filling operations in difficult flow conditions of fly ash – water mixtures

Grzegorz Strozik

Faculty of Mining and Geology, Silesian University of Technology, Akademicka 2, 44-100 Gliwice, Poland

Grzegorz.Strozik@polsl.pl

Abstract. Filling the underground voids created during mining extraction of minerals plays a crucial role in different phases of mining operations. A common practice in Polish coal mines is filling the cavings resulted from coal extraction in longwalls for various safety and technical purposes. Moreover, liquidation of redundant underground workings and decommissioning of mines also require filling of large volumes of voids. As the obsolete mining systems with hydraulic backfill have been entirely replaced by longwalls with caving, a need arose for flexible and easily available method for filing of voids. Nowadays, the most frequently applied are mixtures of fly ash and water, which are able to meet a wide range of requirements of specific applications. These objectives may be achieved by the selection of the type of fly ash and adjustment of concentration of solid in mixture in the respect to geometrical parameters of the transport pipeline. These correlations have been demonstrated on the example of measurements conducted on a laboratory pipeline loop for a wide range of fly ash – water mixtures and flow conditions. Measurement results revealed a complex relationship between rheological properties of mixtures and geometrical parameters of pipelines, which strongly influence the distance at which a mixture may be delivered.

1. Introduction

Handling of voids being left after extraction of minerals and surrounding waste rock belongs to the core of underground mining activities. With the exception for stable main access roads and mine workings in durable hard rocks, most of the mine voids undergo backfilling or are being intentionally left for spontaneously occurring process of rocks breaking and crushing accompanied by subsequent creation and compaction of voids. Such a voids, referred in underground mining of coal as cavings or gob, generate several safety and other adverse issues, which must be addressed by mining engineering [1, 2]

Numerous coal fired power plants distributed all over the area of Upper Silesia Coal basins produce valuable solid waste and by-products, which are easy accessible for coal mines in large quantities and at small distances, what makes them convenient for a bulk us as raw material for preparation of different types of fill mixture. The most interesting among a range of coal combustion by-products and waste is fly ash, in particular the types, which due to containing desulphurization by-product, exhibit significant binding properties in mixture with water [3].

Fill mixtures prepared on the basis of fly with binding properties have found wide range of application in coal mines such as filling of cavings, solidifying backfill, filling of redundant mine workings, underground fire control and extinguishing, reduction of methane emission, spontaneous coal combustion mitigation, ground waters flow control, and many other.

Variability of fly ash physical and chemical properties as well as requirements resulted from application conditions, create a necessity for preparation of mixtures, which exhibit widely differentiated flow properties [4]. In the face of the fact that the hydraulic transport systems in underground coal mines run only by means of gravity, correct assessment and anticipating of mixtures flow parameters is a key factor in successful fill operations.

2. Material and research method

A set of mixtures has been prepared using sample of an average, widely available, and commonly used by the mines type of fly ash. Its maximal grain size is 100 µm and 92% of grains pass through a sieve with 63 µm mesh. Hydraulic density of this fly ash is 2118 kg/m^3. Selected parameters of mixtures used in measurements of flow conditions are listed in table 1.

Range of mixtures densities and concentration of solid particles reflects the full range of their application in underground mining technologies. In rough estimation samples No. 1 – 3 represent mixtures used for underground disposal of salt mine waters, No. 4 – 6 exemplify typical mixtures for filling of cavings, mixtures No. 7 - 8 might be used for filling of redundant mine workings, and No. 9 – 10 may be proposed as solidifying backfill mixtures. Both mass and volume concentrations have been shown while in hydraulic calculations volumetric concentration is in common use, however mixture preparation is triggered rather by mass concentration and water to solids ratio.

Table 1. Parameters of fine-grained mixtures used in laboratory measurements [5].

Number of sample	Density ρ_m (kg/m^3)	Volume Concentration C_v (-)	Mass Concentration C_m (-)	W/S ratio (-)	Unit mass of solids*) u_{cs} (kg/m^3)
1	1105	0.094	0.180	4.555	198.9
2	1209	0.187	0.327	2.053	395.9
3	1277	0.248	0.411	1.433	524.8
4	1307	0.275	0.445	1.247	581.6
5	1382	0.342	0.524	0.910	723.7
6	1447	0.400	0.585	0.709	846.8
7	1486	0.435	0.620	0.614	920.1
8	1529	0.473	0.655	0.526	1002.2
9	1559	0.500	0.679	0.472	1059.0
10	1605	0.541	0.714	0.400	1146.1

*) mass of fly ash per 1 m^3 of mixture

Flow parameters of the mixtures have been measured in a test loop, where specific head loss Δp in kPa/m against velocity of flow data sets have been collected for each mixture. Pipes of diameter 100 mm make a closed loop equipped with a pump, which allow to generate maximal flow velocity v_m of around 5 m/s. Differential measure gauge registers head loss of a flow in horizontal section of a pipe at a distance of two meters. Ultrasonic flowmeter and densometer supply additional data. About 100 dm^3 of mixture circulates continuously in the loop in aim to ensure good mixing of the mixtures compounds and stable flow conditions.

3. Results and analysis

For the reason of readability figure 1 presents results of measurements obtained only for mixtures of densities 1447 kg/m^3 and higher, which are more interesting due to their complex rheological behaviour.

First of all the research have revealed that fly ash - water mixtures under tests behaved as quasi-homogenous ones, what was already predicted theoretically on the basis of grain-size distribution of the fly ash and average free fall velocities of grains of different sizes. On that basis flow of mixtures may be described by rheological models for actual liquids, what reduces amount of empirical formulas staying behind theories of flow of two-phase mixtures in conduits.

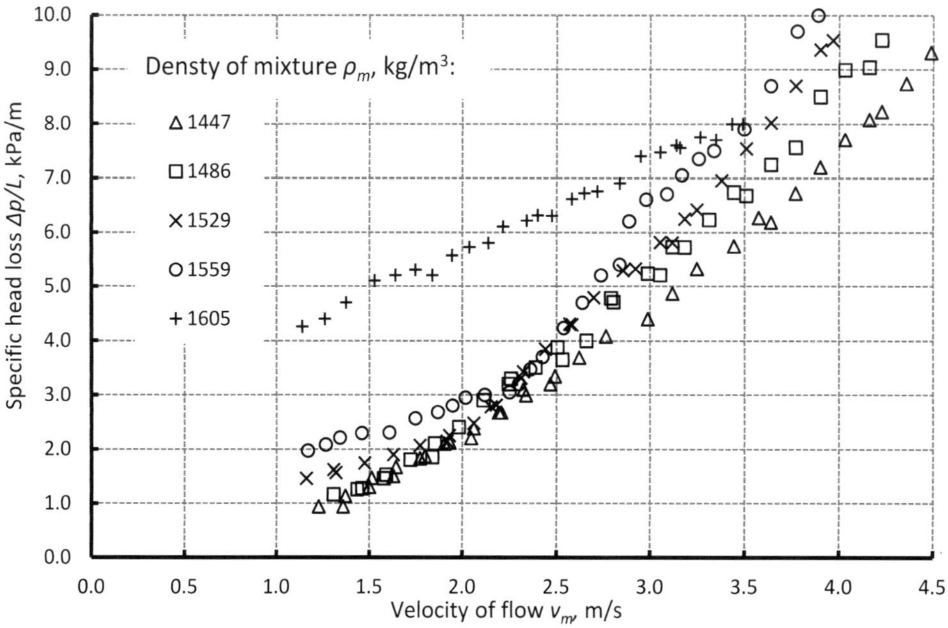

Figure 1. Measured relationship between velocity of flow and specific head loss for fly ash – water mixtures of density from 1447 kg/m^3 to 1605 kg/m^3, pipe diameter D = 80 mm [5].

Flow of liquids is described by the well-known Darcy-Weisbach formula [6], which allows to determine specific head loss in terms of velocity of flow v_m and Darcy friction factor λ:

$$\frac{\Delta p}{L} = \lambda \frac{\rho_m v_m^2}{2D} \tag{1}$$

where Δp is the total head loss of flow in a pipe of length L and diameter D. Friction factor λ is a function of Reynolds number (Re), which depends again on flow velocity, pipe diameter, and density of a liquid, as well as its rheological parameters.

Due to versatility in terms of the use of different rheological models a generalized Reynolds number formula has been applied in a form [7]:

$$Re = \frac{\rho_m v_m^{2-n} D^n}{\frac{\tau_0}{8}\left(\frac{D}{v_m}\right)^n + K\left(\frac{3m+1}{4m}\right)^n 8^{n-1}} \tag{2}$$

where:

$$m = \frac{nK\left(\frac{8v_m}{D}\right)^n}{\tau_0 + K\left(\frac{8v_m}{D}\right)^n} \tag{3}$$

which is adequate for a wide range of rheological flow models. If taking under consideration that structure index n could be equal to 1, than consistency index K is adequate to Bingham dynamic viscosity μ_B. A liquid or mixture, where $n = 1$ and $\tau_0 = 0$ represents properties of a Newtonian fluid.

When the value of *Re* is below about 2300 mixtures flow is laminar and the value of friction factor λ is [8]:

$$\lambda = \frac{64}{Re} \tag{4}$$

For turbulent flows with $Re > \approx$ 4000 exists a wide range of equations allowing determination the friction factor coefficient λ, both for hydraulically smooth and rough pipelines [9]. After statistical assessment of obtained measurement results, for the purposes of further pipe loop tests analysis, empirical equation formulated by Blasius has been chosen:

$$\lambda = \frac{0{,}316}{Re^{0{,}25}} \tag{5}$$

Which is valid for hydraulically smooth pipes, what is the case for the laboratory pipe loop used for the flow tests.

Research works have revealed that mixtures of densities up to ρ_m = 1382 kg/m^3 exhibit rheological properties of Newtonian fluids, while denser mixtures with high accuracy may be considered as Bingham liquids. Values of Bingham model parameters for all mixtures, which took a part in measurements have been listed in table 2.

Values of dynamic viscosity for Newtonian mixtures range from μ_m = 0.010 Pa·s for mixture of density ρ_m = 1105 kg/m^3 up to μ_m = 0.0173 Pa·s for the mixture of density ρ_m = 1382 kg/m^3 with correlation coefficient no less than R^2 = 0.9977. Flow of fly ash - water mixtures of such a low viscosities does not create any substantial problems in the most of typical underground pipeline routes arrangements.

More interesting from the point of the view of mining practice are the rheological parameters of denser mixtures, where yield point appears and coefficients of viscosity (Bingham) may reach substantial values, as presented in table 2.

Table 2. Parameters of fine-grained mixtures used in laboratory measurements [5].

Number of sample	Density ρ_m [kg/m^3]	Yield point τ_0 [Pa]	Bingham Viscosity μ_B [Pa·s]	R^2 [-]
1	1105	0.000	0.0101	0.9982
2	1209	0.000	0.0117	0.9989
3	1277	0.000	0.0149	0.9991
4	1307	0.000	0.0153	0.9977
5	1382	0.000	0.0173	0.9992
6	1447	0.589	0.0313	0.9994
7	1486	1.994	0.0469	0.9954
8	1529	3.634	0.0606	0.9895
9	1559	7.644	0.0807	0.9913
10	1605	24.542	0.1256	0.9963

O the basis of data listed in tables 1 and 2, a relation between volume concentration of a mixture and its rheological parameters may be formulated. Graphs on figure 2 shows that linear increasing of mixture's concentration is accompanying by exponential growth of viscosity and yield point, especially when the concentration of mixture exceeds a value of about 0.47. Above this value the yield point and Bingham viscosity is growing proportionally to the mixtures concentration by the power of 7 and even by the power of 15 for higher concentrations in case of viscosity coefficient. The use of such mixtures in gravitational pipelines in coal mines is debatable and must be very carefully considered.

For given values of mixture's density, viscosity coefficient, yield point, and pipe diameter, flow velocity can be replaced by adequate Reynolds number (*Re*) value, which generally describes actual flow regime which governs the movement of mixtures in the pipe.

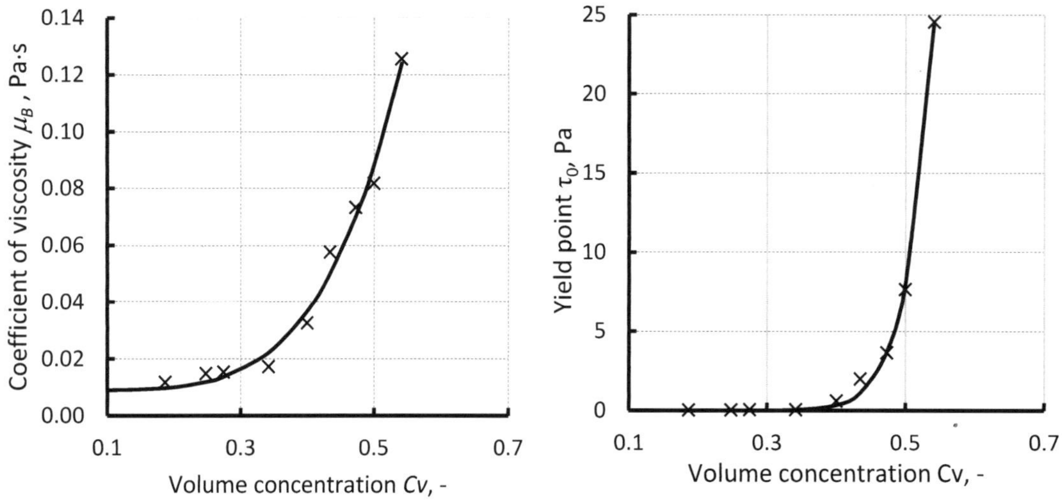

Figure 2. Influence of volume concentration of fly ash – water mixtures under test conditions on coefficient viscosity μ_B on the left and the yield point τ_0 on the right [5].

Range of *Re* values from round 2100 to 4000 represents a gap between laminar and turbulent flow [10, 11] where the head loss cannot be described using friction factor values derived from equations (4) and (5). Some part of data points have been collected under flow conditions of transient regime, enabling the assessment of friction factor variability in the accessible range of mixtures concentrations and flow rates.

Measurements results obtained for mixtures of density in a range of 1486 kg/m^3 ÷ 1559 kg/m^3 show that within the gap of transient flow regime, part of the data points with higher values of *Re* does still follow the turbulent flow dependence, while the rest of the data points measured closer to laminar range of flow follow other, more steep function. This function connects highest data points of laminar flow conditions with lowest data points which fit into the turbulent flow conditions. This observation is clearly visible in measurement results for the mixture of density 1559 kg/m^3, which has been selected as an example plotted in figure 3 for the purposes of this paper.

After statistical analysis of the usefulness of a number of equations available in literature, equation of Blasius (5) has been selected as a basic relation, which was finally rebuild to a form:

$$\lambda = \frac{0{,}011(\rho_m - \rho_w)}{\rho_w Re^{0{,}25}} \tag{6}$$

Range of application of dependence (6) is limited to an area of *Re* values from round 2000 to 2400. Precisely range of this area within transient zone limits has been determined using maximum of correlation coefficients calculated for dependences (4 – 6) for different ranges of measurement points.

Similar impact of Reynolds number value on specific head loss have been observed by the flow of mixtures of lower density (down to 1486 kg/m^3), however area where equation (6) fits the best into the experimental data becomes narrower with decreasing density of a mixture and is gradually replaced by the turbulent flow approximation. In case of mixture of density 1605 kg/m^3, which was the highest in the range of research, full range of measured flows was laminar, so there is no evidence about suspected extent of transient area and the validity range of equation (6).

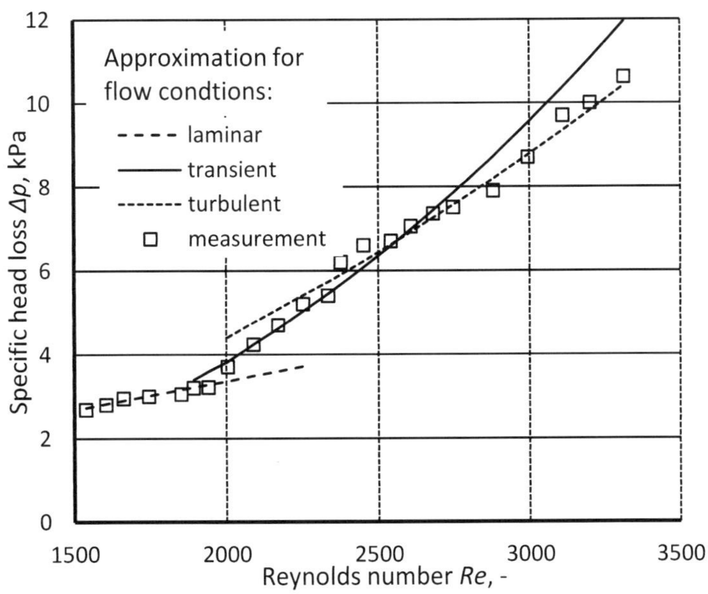

Figure 3. Specific head loss in a function of Reynolds number for mixture of density 1555 kg/m^3. approximated with equations (4) for laminar flow, (5) for turbulent flow regime, and (6) for the transient area of flow [5].

4. Practical application of the research results in gravitational pipeline transport of fly ash – water mixtures

In gravitational hydraulic transport systems the only source of energy is the maximum head of mixture resulted from difference in height between inlet (H_i) and outlet of a pipeline (H_o). If the mixture preparation device is able to keep the mixture level at the top of the pipeline inlet, than the pressure, which converts into head loss during flow of a mixture equals:

$$\Delta p = \rho_m g \Delta (H_i - H_o) \tag{7}$$

Taking under considerations the relationship between properties of mixture and geometrical shape of pipeline described in equations (1) and (7), a mixture of certain density and rheological parameters is able to flow at a determined flow rate for a limited distance.

While the depth of a mine, which corresponds to the pipeline height difference is more or less constant, thus the horizontal distances between mine shafts and places of operations can be substantial, frequently even more than twenty times longer as the length of vertical part of a pipeline.

Large number of analyses of underground voids fill processes show that in most cases the flow occurs in the range of Reynolds number values from round 2000 up to about 12000, so the flow tests in laboratory test loop reflect directly the conditions governing operation of industrial pipelines, with respect to the difference in pipe diameters (80 mm in the laboratory loop and 100 – 150 mm in underground pipeline systems in coal mines).

The main risk for the effective and safe operation of pipeline systems for filling of voids is blocking of the flow in pipe resulting from too long transport way. In practice mine pipelines must reach certain, often relatively distant areas in an underground mine and deliver fill mixture of expected parameters with acceptable flow rate. Flow rate value requirements may be resulting from technical limitations like time of mining cycle or properties of mixture, i.e. time of hydration (binding process). From the other side, eventual underestimation of hydraulic transport capacity may compromise safety issues and generate increase of mining operations costs.

Diagram on figure 4 anticipates distance of effective transport with a flow rate Q = 100 m^3/h in a pipe of diameter of D = 100 mm depending on density of mixture and flow regime. In the first place is to be seen how the transport distance varies with changing mixture's density. Increase of density leads to less intensive decrease of transport distance in turbulent than in laminar flow zone. Unavoidable is the threshold between turbulent and laminar flow zone, so in case depicted in Fig. 4 would be advantageous to omit densities of mixture in the range of about 1600 kg/m^3 up to 1615 kg/m^3.

Although basically increasing density/decreasing transport distance relation is trivial one, attention should be given to an effect of mixture preparation inaccuracies on the flow distance change. Considering situation shown in figure 4, it can be found that for example change of mixture's density from 1486 kg/m^3 to 1529 kg/m^3 (by only 2.89%), what may be referred to about 5.6% mass concentration volatility in mixture preparation plant, finally leads to about 12.6% reduction of transport length (by 174 m). In laminar flow zone this correlation is even stronger and for example, in case of 1.23% increase of density from 1630 to 1650 kg/m^3 (round 2% increase of mass concentration) results in almost 24% reduction of transport length from 1150 down to 876 m. Such a variability of critical parameter, which is the range of effective transport may not only reduce significantly the flow rate at which a fill mixture is being delivered to the place of application, but also may lead to a failure of hydraulic transport system. There is a need to keep precisely the composition of fly ash – water mixture, what may be a challenge for the mixture preparation plant.

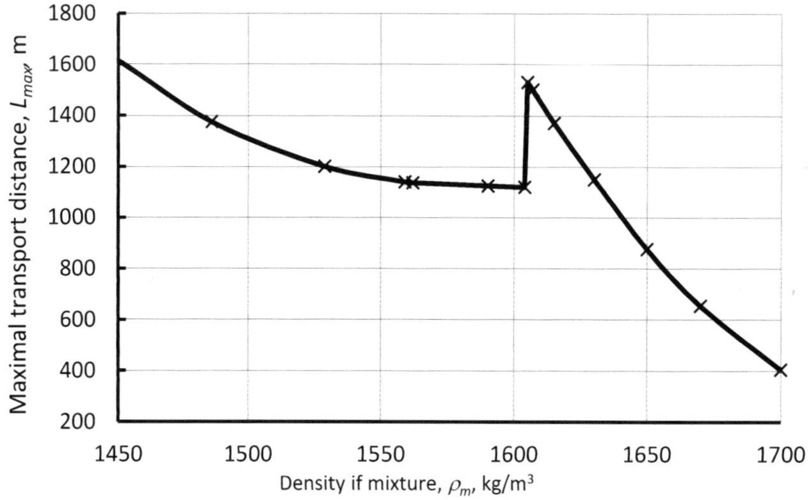

Figure 4. Maximal transport distance in relation to the density of mixture in a pipeline of height difference 300 m and pipe diameter 100 mm.

5. Conclusion

Measurements of flow properties of fly ash – water mixtures used in underground mine operations has been conducted on a laboratory test loop, however achieved flow parameters are very close to actual conditions occurring in hydraulic pipeline transport systems in coal mines.

Results of research and their analysis discussed in the paper represent only a part of the whole research work. The attention has been given to one of the priority goals in design of underground filling of voids, namely range of transport of fly ash – water mixtures and the impact of flow conditions and rheological properties of mixtures on its variability.

References

[1] Popczyk M 2017 *IOP Conf. Ser.: Mater. Sci. Eng.* 268 012012
[2] Palarski J and Zając A 2017 The future of backfill in underground coal mines - underground placement of fill material based on fly ash and coal processing waste *12th International Conference on Mining with Backfill 2017* Englewood: Society for Mining, Metallurgy and Exploration (Colorado: Denver) pp 254–61

[3] Pierzyna P and Popczyk M 2014 Odzysk odpadu energetycznego z metody mokrego odsiarczania spalin do likwidacji zbędnych wyrobisk korytarzowych *Polityka Energetyczna – Energy Policy Journal* **17(3)** pp 341–354
[4] Plewa F, Popczyk M and Pierzyna P 2013 Możliwości wykorzystania wybranych odpadów energetycznych z udziałem środka wiążącego do podsadzki zestalanej w podziemiu kopalń *Polityka Energetyczna – Energy Policy Journa* **16(4)** pp 257–270
[5] Strozik G 2018 *Wybrane zagadnienia transportu i zastosowania hydromieszanin drobnorakcyjnych produktów spalania węgla kamiennego w górnictwie podziemnym* (Gliwice: Wydawnictwo Politechniki Śląskiej)
[6] Albrigth L 2008 *Albright's Chemical Engineering Handbook* (Boca Raton: CRC Press)
[7] Madlener K, Frey B and Ciezki H K 2009 Generalized Reynolds Number for Non-Newtonian Fluids *Progress in Propulsion Physics* 1 pp 237-250
[8] Puzyrewski J and Sawicki J 2016 *Podstawy mechaniki płynów i hydrauliki* (Warszawa: Wydawnictwo Naukowe PWN)
[9] Mays L W 2004 *Hydraulic design handbook* (New York: The McGraw-Hill Companies)
[10] Eckhardt B 2012 Turbulence transition in shear flows: chaos in high-dimensional spaces. *Procedia IUTAM* 5 pp 165–168
[11] Kernswell R R 2005 Recent Progress in Understanding The Transition Threshold to Turbulence in a Pipe *Phys. Rev. Lett.* 91 pp 17–44

IOP Conf. Series: Earth and Environmental Science **174** (2018) 012015 doi:10.1088/1755-1315/174/1/012015

Influence of pipe roughness and coating built-up on pipe walls on the flow of solidifying Non-newtonian fly ash-water mixtures in hydraulic transport systems in coal mines

Grzegorz Strozik

Faculty of Mining and Geology, Silesian University of Technology, Akademicka 2, 44-100 Gliwice, Poland

grzegorz.strozik@polsl.pl

Abstract. Due to favourable physical properties, the most frequently used fill materials in underground technologies are mixtures of fly ash from hard coal combustion and water. The mixtures are delivered by pipelines from the ground surface to the places of application gravitationally, so the reliable assessment of flow parameters is crucial for the successful fill operations. In first place rheological properties of mixtures must be known for further flow parameters prediction. These parameters may be obtained from measurements in laboratory pipeline loop, which reflect the operation conditions of industrial pipelines in terms of flow rate range and variability of rheological properties of mixtures. Further considerations focused on the problems of range of flow and influence of pipe roughness on reduction of mixture delivery distance. Among other, the conclusion is that the transport distance occurring with technologically acceptable flow rate decreases rapidly with increasing pipe roughness and may finally lead to pipeline failure. As it has been shown on an example to any point of mining operations may be delivered fill mixtures with a certain maximal value. This maximal density is varying widely in relation to actual length and height of gravitational pipeline. Flow parameters of fill mixtures in gravitational pipelines may be affected heavily by increasing of pipe roughness. Variability of absolute pipe roughness have been discussed in relation to typical abrasion and corrosion processes of pipe walls and also considering potential coating built-up created by solidified mixture layers. Operation of gravitational pipeline in systems of underground voids filing should be considered as a dynamic process, where the geometrical parameters and roughness of the pipeline may undergo continuous and rapid changes.

1. Introduction

Underground hard coal mines in Poland are practicing widely a lot of quasi-backfill operations with use of fly ash from hard coal combustion. Filling voids accompanying longwall system with caving is the most frequently adopted technology aimed on spontaneous coal combustion mitigation and improvement of ventilation conditions in longwalls. Approximately in 30 from about 80 currently running longwalls this technology is adopted and allows to safe and efficient coal production. Coal mines also use fly ash – water mixtures for other purposes, as for example liquidation of abandoned workings or construction of backfill plugs. The wide range of applications means that fly ash – water mixtures with different characteristics must be produced on the surface and delivered underground.

Growing distances between shafts (mine centers) and mining operations areas create increasingly challenging conditions for hydraulic transportation of fill mixtures. A research program has undertaken to analyze the influence of concentration of a fly ash in mixture with water on its rheological properties, which were measured in laboratory pipeline loop. Fly ash from a power plant combustion vessel with semi-dry desulphurization method was selected as an adequately representative example from the range of coal combustion solid waste and by-products available on local market, however it should be kept in mind that each of such substances may exhibit different rheological and solidifying characteristics in mixture with water [1].

2. Measurement of fly ash – water mixture rheological properties in laboratory pipe loop

Head loss/flow velocity relations have been measured for fly ash – water mixtures in the range of density from 1105 to 1605 kg/m^3 circulating in a pipe loop of diameter 80 mm, which was able to keep a flow velocity from around 1.0 m/s up to 4.5 m/s. Analysis of measurement results have been conducted assuming homogeneity of fly ash water mixtures and constant value of rheological properties during the time of flow in the pipeline. First assumption was discussed on the basis of the classification of mixtures concerning existence of pseudo homogenous, quasi-homogenous and heterogonous mixtures [2] with respect to their dispersity [3].

The second assumption is justified by the time differences between flow time in mine pipelines systems (maximally up to about half an hour in the state of continuous movement counteracting binding) and initiation of binding (more than 2 hours in an immobilized state).

For the purposes of further considerations basic parameters of fly ash – water mixtures together with their rheological properties accordingly to Bingham plastic model have been listed in table 1. Mixture of densities smaller than about 1400 kg/m^3 are Newtonian fluids ($\tau_0 = 0$ Pa)

Remarkable is strong increase of yield point and Bingham viscosity coefficient, which are in exponential relation with density and concentration of a mixture. From the point of the view of production of fly ash-water mixtures in mine fill preparation plants, an important conclusion from the analyse of the data in table 1 is that even small relative fluctuations in water to solid material ratio (W/S) result in high variability of rheological parameters. For example a 10% decrease of W/S ratio from 0.526 to 0.472 increases the yield point of the mixture by 110%, from 3.634 up to 7.644 Pa.

Table 1. Parameters of fine-grained mixtures used in laboratory measurements [4].

Number of sample	Density ρ_m (kg/m^3)	Yield point τ_0 (Pa)	Bingham Viscosity μ_B (Pa·s)	Volume Concentration C_v (-)	W/S ratio (-)
1	1382	0.000	0.0173	0.342	0.910
2	1447	0.589	0.0313	0.400	0.709
3	1486	1.994	0.0469	0.435	0.614
4	1529	3.634	0.0606	0.473	0.526
5	1559	7.644	0.0807	0.500	0.472
6	1605	24.542	0.1256	0.541	0.400

*) mass of fly ash per 1 m^3 of mixture

3. Gravitational flow of mixtures in pipes

To ensure successful operation of a longwall it is necessary to be able to deliver the fill mixture at the full length of longwall panels and to each point of an accessed part of a coal seam. This requirement, however, could be difficult to meet due to considerable dimensions of longwall panels and large distance between the mine shafts and areas of mining operations.

Flow of a liquid (either homogenous mixture) in a pipeline is governed by Darcy-Weisbach equation [5, 6]:

$$\Delta p = \lambda \frac{L}{D} \frac{\rho_m v_m^2}{2} \text{ (Pa)} \tag{1}$$

where: L – length of pipeline, m,
D – pipe diameter, m,

v_m^2 – velocity of flow, m/s
λ – Darcy friction factor, -.
Δp – total head loss, Pa.

Darcy friction factor λ depends on Reynolds number of the flow, which in turn is related to flow velocity, density of the liquid (homogenous mixture) and its rheological properties. The main area of interest of the above mentioned research work was the assessment of Darcy friction factor for flows of fly ash – water mixtures under different flow conditions (laminar, turbulent and transient zone) in a wide range of fly ash concentration in mixtures.

In gravitational pipeline transport systems the only source of energy is potential energy of a mixture represented by its hydrostatic pressure resulted from level difference between inlet and outlet of the pipeline. This pressure is also formulated as disposable pressure and must be balanced by the head loss:

$$\Delta p = \rho_m g \Delta H \text{ (Pa)} \tag{2}$$

For the analyses of flow in rough pipes in turbulent zone the commonly used equation of Colebrook-White has been adopted [7, 8]:

$$\frac{1}{\sqrt{\lambda}} = -2\log\left[\frac{2{,}51}{\sqrt{\lambda}\,Re} + \frac{\varepsilon}{3{,}71D}\right] \text{ (Pa)} \tag{3}$$

where: ε – absolute pipe roughness, m.

Equation (3) is valid in the area of turbulent flow with laminar sublayer of the thickness smaller than the roughness of the pipe wall, so the friction factor λ depends both on Reynolds number and pipe roughness. This formulation is valid for relatively low range of *Re*, from round 4×10^3 up to about 10^5 [9]. By increasing value of *Re* the fraction containing it may be neglected and the Darcy friction factor depends only on pipe roughness in highly turbulent flows [7].

Practice shows that mine gravitational pipelines do not operate at Reynolds number high enough to reach the fully developed turbulent flow, where the friction factor depends only on pipe wall roughness. Also laminar flow is not a typical operational mode in gravitational pipelines. Observations of mine gravitational pipelines operation have shown that in most cases the flow of mixtures occurs at Reynolds number values in a range from about 3000 up to 12000 and therefore in transient zone and not fully-developed turbulent flow regime, where the influence of pipe roughness on the Darcy friction factor is substantial. Similar is the range of the Reynolds number in laboratory tests of fly ash – mixture flow, varied from about 1000 do 30000. Comparability od Reynolds number is one of the similarity criteria for flows.

4. Analysis of transport range in mine gravitational pipelines

Each particular system of gravitational hydraulic transport can be described by its length, diameter, and height (elevation difference between inlet and outlet of a pipeline) [10]. In case of hydraulic systems for filling underground voids in longwall cavings the situation is much more complex, while both the length and height of a pipeline are variable. Considering simplified example of a longwall mining layout shown in figure 1 the main route of pipeline from point P0 (mixture preparation plant) on the surface to the point P1 is 2880 m long and the elevation difference is 750 m. During operation of each longwall from L1 to L4, fly ash – water mixture must be delivered to each point of current position of a front of a longwall. So the pipelines must reach points from EP1 to EP 4 in longwalls L1 to L4 respectively. In such a case variability of length and height of pipeline can be described for each longwall as follows:

- Longwall 1: length L_{L1} from 3380 to 4580 m, $\Delta H_{L1} = 690$ m,
- Longwall 2: length L_{L2} from 3200 to 4350 m, $\Delta H_{L2} = 710$ m,
- Longwall 3: length L_{L3} from 3040 to 4040 m, $\Delta H_{L3} = 730$ m,
- Longwall 4: length L_{L3} from 2880 to 3660 m, $\Delta H_{L3} = 750$ m.

Transport range in the context of delivery of fly ash –water mixtures to longwalls for the purposes of filling of voids in underground mines is such a maximal length of an gravitational pipeline, which in given conditions of pipeline height and mixture density ensures at least such a minimal flow rate, which is acceptable from any adopted technological requirements. In this case a flow rate of Q_m = 100 m^3/h has been chosen, because most of existing mixture preparation plants are equipped with mixers of such a capacity. In a pipeline of a constant diameter D = 100 mm it results in flow velocity v_m = 3.538 m/s. Relevant Reynolds number values accompanying the flow are varying from 2427 for the mixture of density 1605 kg/m^3 up to 28327 for the mixture of density 1382 kg/m^3. Lengths of pipelines (transport ranges) at which the flow occurs with the velocity v_m =3.538 m/s have been listed in table 2.

As it can be seen, possibility for the use of the mixture of the highest density is very limited, and is possible only in parts of longwalls L1 and L2 where the pipelines in horizontal gates are short. Remarkable observation is that the use of mixture of density 1605 kg/m^3 is not possible when the roughness of the pipeline have been taken under consideration.

Permissible maximal density of fill mixture is subsequently decreasing in more distant parts of the longwall panels L1 ÷ L4. During mining of these longwalls, the current mixture properties should be constantly adjusted in aim to achieve best possible effects of filling.

Table 2. Transport range of fly ash water – mixtures in relation to density of mixture, pipe wall roughness and elevation difference. Flow rate Q_m = 100 m^3/h, pipe diameter D = 100 mm [4]

Roughness ε [mm]	Density ρ_m [kg/m^3]				
	1382	1447	1529	1559	1605
			ΔH = 690 m		
0	6660	4676	4374	3763	2936
1	3973	3392	3338	2976	2516
3	2724	2446	2463	2240	2010
			ΔH = 710 m		
0	6853	4812	4500	3872	3021
1	4088	3490	3434	3063	2589
3	2804	2517	2533	2306	2069
			ΔH = 730 m		
0	6730	4948	4627	3982	3107
1	4015	3590	3531	3150	2661
3	2754	2587	2605	2370	2127
			ΔH = 750 m		
0	7238	5083	4754	4091	3191
1	4319	3688	3627	3235	2734
3	2962	2659	2677	2436	2184

For the conditions of discussed example, considering the pipeline as hydraulically smooth, the range of transport is changing from 2936 m to 6660 in dependence to density of mixture. Distribution of maximal mixture's density over the area of exemplary mining operations are shown in figure 1. As it can be seen from the figures, shortening of the pipeline allows to use thicker (with higher water to solids ratio) fly ash – water mixtures, which is beneficial for the results of the filling of underground voids. Alternatively, the picture shows maximal densities of mixture, which exceeding may lead to a failure of the transport system. To avoid such a risk, reliable control of mixture parameters in mixture preparation plant, precise analysis and determination of gravitational pipelines work parameters, and regular measurements of rheological parameters of fly ash – water mixture must be ensured. The last one is often an underestimated risk factor for the operation of mine systems of voids filling. Rheological measurements conducted on regular basis show that the physical properties of fly ash may undergo substantial variability from one delivery to another, even if they come from the same source.

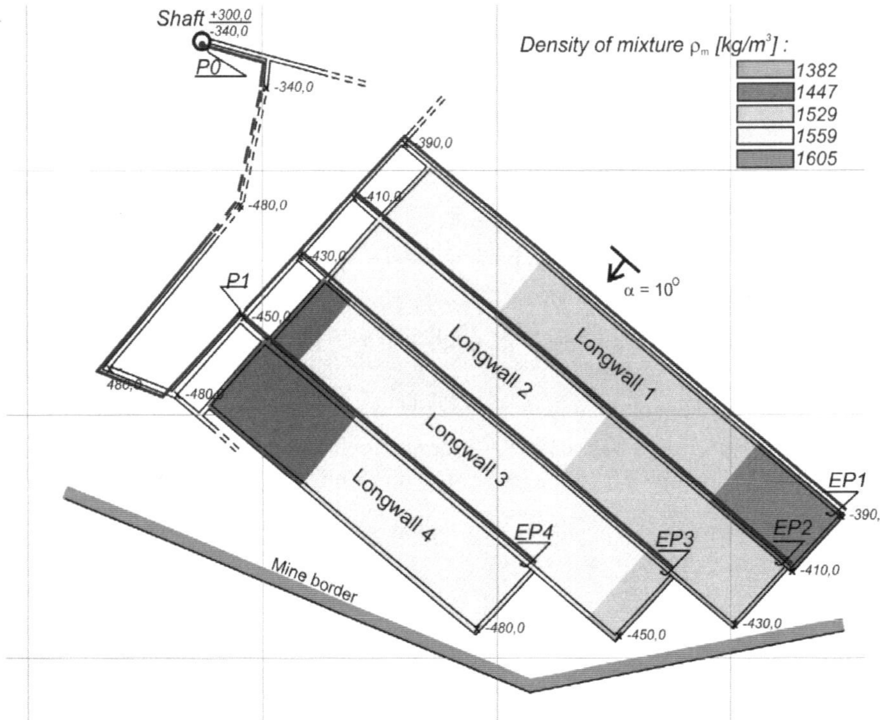

Figure 1. Filling of cavings in a panel of longwalls – distribution of maximal density of fill mixture calculated for hydraulically smooth pipeline (see text for details).

5. Roughness of pipelines for hydraulic transport of mine fill mixtures

Roughness of pipes in mine gravitational transport systems can be significantly differentiated depending on the type of pipes used for the pipeline assembly, their technical condition, and degree of wear. Moreover, pipe roughness can increase along with the time of pipeline's lifetime due to corrosion, wear, and – in case of mixtures which exhibit binding properties – build-up of a coating formed from solidified layers of fly ash –waters mixture. Although roughness of new steel pipes is less than 0.1 mm, used and slightly corroded pipes may exhibit absolute roughness of about 0.4 to 1.0 mm. In case of cast iron pipes, even relatively insignificantly eroded pipe surfaces have roughness of more than 1.5 mm [10].

During a flow of fly ash – water mixture in laminar mode or in turbulent mode with laminar sub-layer, what is a case by low values of Reynolds number as observed in mine pipelines, a thin stationary film of a mixture covers pipe walls and may be able to solidify if there is no or incorrect rinsing of the pipeline after cycle of transport. As the fill materials after solidification often do not undergo deterioration in the presence of water, subsequent mixture flow cycles may lead to the build-up of rough and durable coating of the pipe walls (mostly at the bottom).

Absolute roughness of such a surface is similar as in case of concrete and its average roughness is about 3 mm. Old, highly eroded, and pipes with severe wall coating built-up may exhibit roughness even of 10 mm [11].

Figure 2 shows how increased pipeline roughness reduces transport capability of a gravitational pipeline. In the same conditions as in the case presented in figure 1, where the flow parameters have been calculated assuming that the pipeline is hydraulically smooth, absolute roughness of 1.0 mm reduces densities of mixture, which may be delivered to any points of considered longwalls. As it can be seen in this example, when the pipeline is hydraulically smooth in the most part of the mining area mixture of density 1559 kg/m^3 may be used (figure 1), although considering rough pipe, the actual density of mixture, which may be used for filling of cavings is only 1447 – 1529 kg/m^3. Also significantly decreased must be density of the fill mixture for filling of voids in the most distant part of longwalls L1 and L2. Larger difference between actual and assumed (or neglected) pipeline roughness results in much more rapid decrease of maximal mixture density values and in practice may lead to immobilization of mixture in pipeline.

Average pipeline roughness may undergo continuous and rapid changes in time. Corrosion and abrasion are processes that take place slowly over time. A rough wall coating may be built up after only a view pipeline operation cycles. Also connecting or disconnecting of a pipeline section may result in rapid change of average roughness of the whole pipeline.

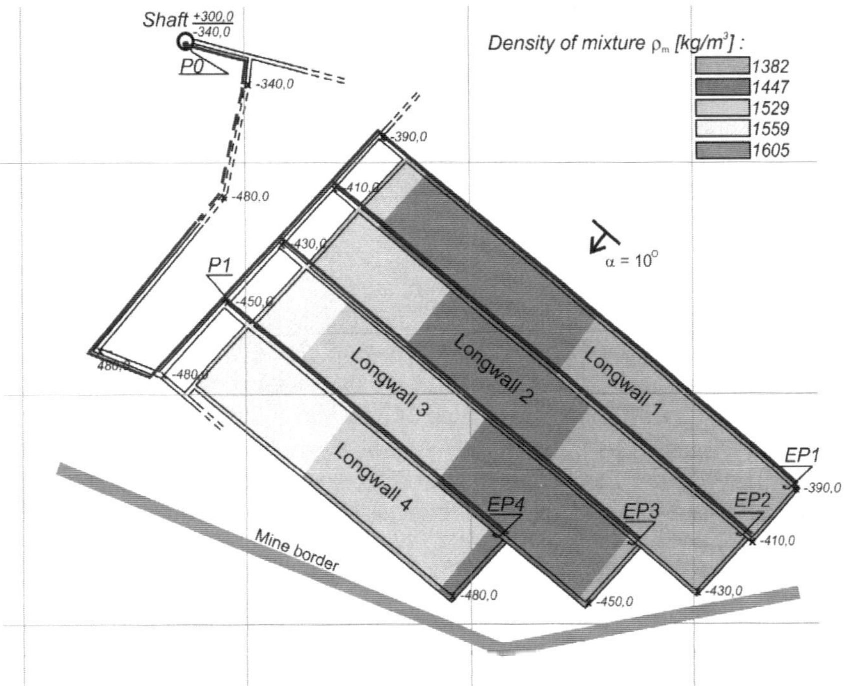

Figure 2. Filling of cavings in a panel of longwalls – distribution of maximal density of fill mixture calculated for rough pipeline (ε = 1.0 mm) (see text for details).

The often encountered situation in mine pipelines is that the output of the pipeline lies at higher altitude as the lowest part of the pipeline. In such a case a non-runoff trough is present in the geometric and also hydraulic profile of the pipeline (figure 3). Mostly there is no possibility to discharge water left after rinsing from the recessed parts of the pipeline and the pipe walls are exposed on increased corrosion effects. Also rinsing of such a pipelines is more difficult than in typical, uniformly inclined conduits, and the conditions for coating built-up process are favourable.

Due to rigorous mine safety regulations as well as difficult operational conditions, until recently plastic pipes had not been implemented in underground coal mines. However UHMWPE polyethylene pipes [12] started to be used, which are highly abrasion and wear resistant and may be used for transport of liquid media under pressure of up to 20 MPa. From the point of the view of hydraulic friction polyethylene pipes maybe considered as hydraulically smooth and offering better flow conditions especially at long distances.

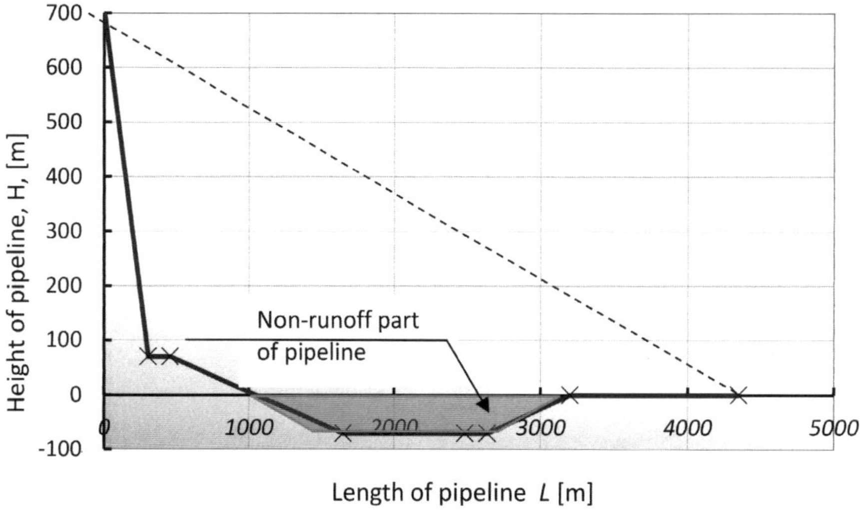

Figure 3. Hydraulic profile of the pipeline to longwall L4 (from P0 to EP1 in figure 1).

6. Conclusion

Hydraulic transport of fly –ash water mixtures in gravitational pipeline systems requires very careful composition of the mixtures because of their highly variable rheological properties. In order to obtain the best results in the filling of voids, fly ash – water mixtures of density in range of about 1450 ÷ 1550 kg/m^3 should be used. Such a mixtures exhibit properties of Non-newtonian fluids and may be characterized by viscosity coefficient in the range of about μ_B = 0.03 ÷ 0.08 Pa·s and yield point of about τ_0 = 0.6 ÷ 6.0 Pa in terms of Bingham (plastic) rheological model parameters. Mixture of such rheological properties require reliable assessment of flow parameters in pipelines to avoid immobilization of the flow. Also mixture preparation plants must be able to eliminate fluctuations of components proportions. In case of highly viscous Non-newtonian mixtures implementation special attention should be given to the evaluation of pipeline roughness. Appropriate incorporation of the pipe walls roughness will allow precise and reliable determination of maximal thickness of fly ash – water mixture and reduce a risk of pipeline transport failure due to blockage of mixture flow in pipe. The optimization process of hydraulic fill technology in a coal mine is a permanent striving to a compromise between performance of a mixture as a fill material and safe operation of gravitational pipeline transport system. An additional difficulty is that in contrary to typical hydraulic transport pipelines, the system is not static and the length and elevation difference of a pipeline depends on its current position in longwall. Also average pipeline roughness maybe subject of continuous or step changes, thus requiring an appropriate design approach.

References

[1] Pierzyna P and Popczyk M 2014 Odzysk odpadu energetycznego z metody mokrego odsiarczania spalin do likwidacji zbędnych wyrobisk korytarzowych. *Polityka Energetyczna – Energy Policy Journal* **17(3)** pp 341-54
[2] Kleinstreuer C 2018 *Modern fluid dynamics* (Boca Raton: CRC Press Taylor & Francis Group)
[3] Buhrke H , Kecke H J and Richter H 1989 *Strömungsförderer* (Berlin: VEB Verlag Technik)
[4] Strozik G 2018 *Wybrane zagadnienia transportu i zastosowania hydromieszanin drobnofrakcyjnych produktów spalania węgla kamiennego w górnictwie podziemnym* (Gliwice: Wydawnictwo Politechniki Śląskiej)
[5] Albrigth L 2008 *Albright's Chemical Engineering Handbook* (Boca Raton: CRC Press Taylor & Francis Group)

[6] Puzyrewski J and Sawicki J 2016 *Podstawy mechaniki płynów i hydrauliki* (Warszawa: Wydawnictwo Naukowe PWN)
[7] Keady G 1998 Colebrook-White formula for pipe flows *Journal of Hydraulic Eng.* **1** pp 96-7
[8] Slatter P T 1999 The role of rheology in the pipelining of mineral slurries *Min. Pro. Ext. Met. Rev.* **20** pp 281-300
[9] Eckhardt B 2012 Turbulence transition in shear flows: chaos in high-dimensional spaces *Procedia IUTAM* **5** pp 165-168
[10] Marcin Popczyk 2017 *IOP Conf. Ser.: Mater. Sci. Eng.* 268 012012
[11] PN-87/M-04250 Surface layer – terminology.
[12] http://slurrypipes.com.au/uhmwpe-pipe/

IOP Conf. Series: Earth and Environmental Science **174** (2018) 012016 doi:10.1088/1755-1315/174/1/012016

Working environment and observance of occupational health and safety regulations – case study

Małgorzata Wyganowska and Katarzyna Tobór-Osadnik

Faculty of Mining and Geology, Silesian University of Technology, Akademicka 2, 44-100, Gliwice, Poland

malgorzata.wyganowska@polsl.pl

Abstract. Safety culture can be seen from many aspects. Some people in their definitions place emphasis on observable indicators of safety culture, while others concentrate more on the identification of basic principles of safety culture, which affect the „safe behaviour" of organisation members. Safety culture seen as a part of organisation's culture is reflected in rules, values, attitudes and behaviour of its employees. One of the areas of the improvement of working environment is the optimisation of workstations in line with ergonomics principles. The areas of the improvement of ergonomic working conditions include: the amount of energy expenditure, static load connected with the work performed, the repeatability of working movements, the type and availability of personal protective equipment. The paper has presented a part of the work safety audit carried out in June 2016 in a mining-related heavy industry company from Silesian voivodeship. The results of the presented survey show that creating appropriate working environment affects occupational health and safety not only in the area of optimisation of protection against hazards, but also building desired employee's behaviour.

1. Introduction

Working environment, which can be defined as: „the entirety of the conditions in which work is carried out", depends on various material (physical) and non-material (social) factors affecting, among others, the safety and health of employees and, consequently, work efficiency and effectiveness" [1].

Today one of the first activities to create suitable working environment is building desired safety culture [2]. The notion of safety culture was first used by a group of consultants of the International Atomic Energy Agency (IAEA) after the nuclear catastrophy in Chernobyl in 1986 [3].

Safety culture can be seen from many aspects. Some people in their definitions place emphasis on observable indicators of safety culture, while others concentrate more on the identification of basic principles of safety culture, which affect the „safe behaviour" of organisation members. Safety culture seen as a part of organisation's culture is reflected in rules, values, attitudes and behaviour of its employees. The definition of safety culture according to the Health and Safety Executive, which seems to reflect well the characteristics of this area of organisation's culture, is as follows: „OHS culture is a result of the individual and group values, attitudes, perception, competence, behaviour patterns as well as style and quality of safety management in an organisation" [4].

Building safety culture and, within its scope, desired working environment uses numerous activities concerning OHS. Such activities usually include, first of all, an appropriate qualitative choice of personal protective equipment, developing and perfecting OHS procedures or improving the ergonomics of workstations and working conditions. To measure working conditions the following are among others used [5, 6, 7]:

- checklists,
- measurements of physical working environment,
- accident reports,
- audits.

Using these tools usually helps to improve current working conditions and, consequently, improves the health and safety of employees in a given environment.

One of the areas of the improvement of working environment is the optimisation of workstations in line with ergonomics principles. The areas of the improvement of ergonomic working conditions include: the amount of energy expenditure, static load connected with the work performed, the repeatability of working movements, the type and availability of personal protective equipment. Appropriately prepared manuals of procedures, and their availability and intelligibility are also an important part of appropriate working conditions and employees' behaviour at their workstations [8].

Physical workload can be measured with the amount of energy expenditure during the entire shift, understood as the sum of the energy used to perform individual activities [9]. The assessment of energy expenditure includes: the assessment of dynamic load, which is expending energy by the body in relation to the effort, the assessment of static load resulting from the position of the body at work, or the assessment of the repeatability of movements.

Workstation instructions prepared in an appropriate manner are not sufficient for the desired behaviour of employees [10]. They should be available and clear and enable continuous improvement of learned procedures. Therefore, the quality of instructions and familiarity with them can also be an important indicator for the assessment of safety culture level, including OHS. They can also be a key factor conditioning employees' safe behaviour and observance of OHS regulations.

2. Description of the survey

As part of the work safety audit carried out in June 2016 in a mining-related heavy industry company from Silesian voivodeship, a survey amongst experts in selected areas of occupational safety management was carried out. An anonymous questionnaire was used as a research tool. The respondents were randomly selected amongst blue-collar/production workers. This article will present a fragment of the above survey, concerning the impact of the selected working environment elements on the observance of OHS regulations. The aim of this fragment of the survey was to define potential areas of irregularities and necessary changes in the process of OHS management in the subject company[11].

The results of the survey showed factors present in the respondents' working environment and affecting the observance of OHS regulations. The respondents assessed whether a given factor was present in their working environment, and if any factor was chosen, they had to rate on the scale from 1 (does not affect) to 5 (affects very significantly) to what extent it affected the observance of OHS regulations and, consequently, safe work.

What is important for further conclusions, it should be noted that in the subject company strong safety culture was diagnosed [3]. For the purposes of assessing a culture level not only was the accident rate taken into account, but also the type and number of OHS training courses, a communication system in this area, indication systems used or workstation tidiness. In the following chapter the results of the survey concerning selected working environment elements will be presented.

3. Overview of survey results

In the opinion of more than half of the respondents physical workload affects the observance of OHS regulation to a large or a very large extent (4.0 and 5.0). The difficulty degree of the work carried out (measured with energy expenditure level), the load resulting from unnatural (forced) positions of the body, as well as the repeatability of the same working movements with simultaneous engagement of limited number of muscle groups may in the respondents' opinion result in violating OHS rules and consequently accidents (or near misses).

In the case of energy expenditure connected with performed work, the problem is recognised by 68% of the respondents (figure 1), in the case of static load by as much as 81% (figure 2), and in the case of the repeatability of the same working movement – 71% of the respondents (figure 3). None of the respondents thought that the amount of workload did not affect the observance of OHS regulations (although in the case of static load 7% of the respondents answered "does not affect", and in the case of the repeatability of the same working movement – the result was 5%).

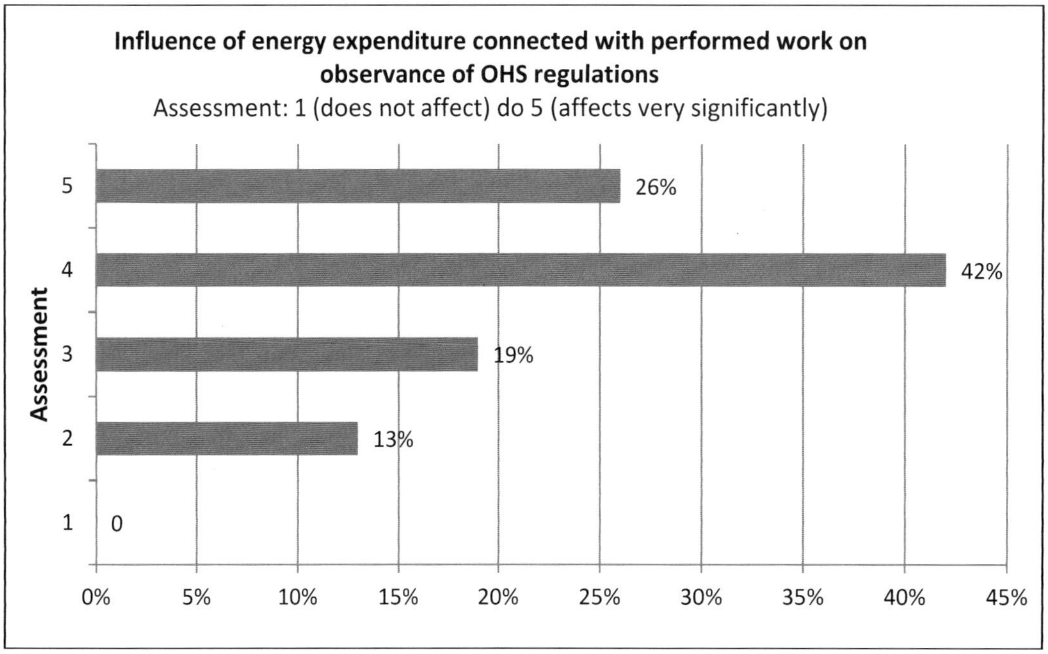

Figure 1. Influence of energy expenditure connected with performed work on observance of OHS regulations.

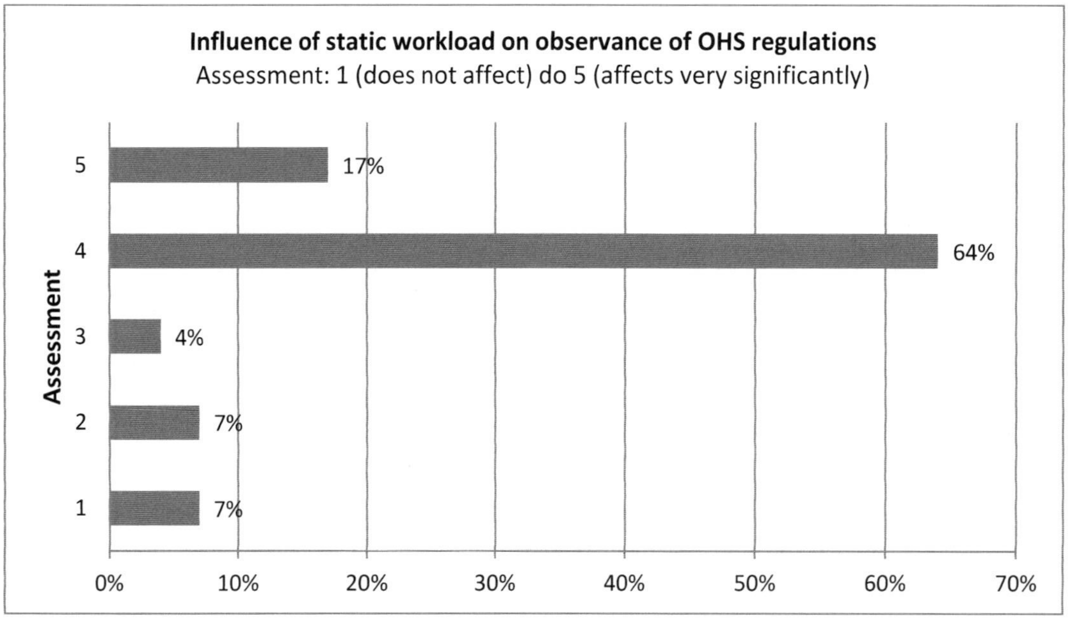

Figure 2. Influence of static workload on observance of OHS regulations.

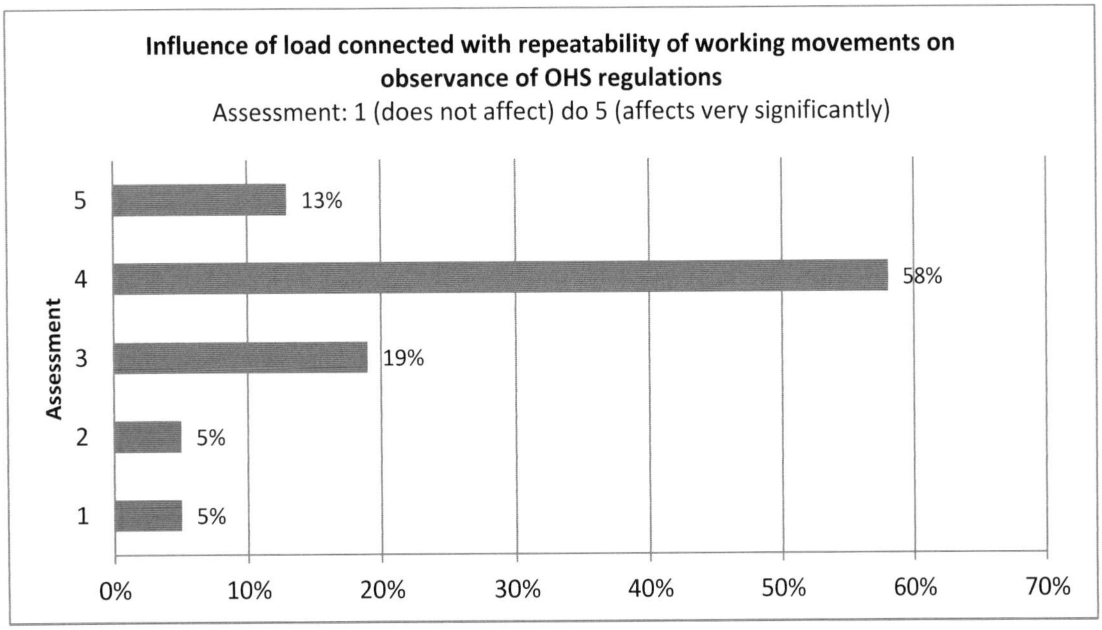

Figure 3. Influence of load connected with repeatability of working movements on observance of OHS regulations.

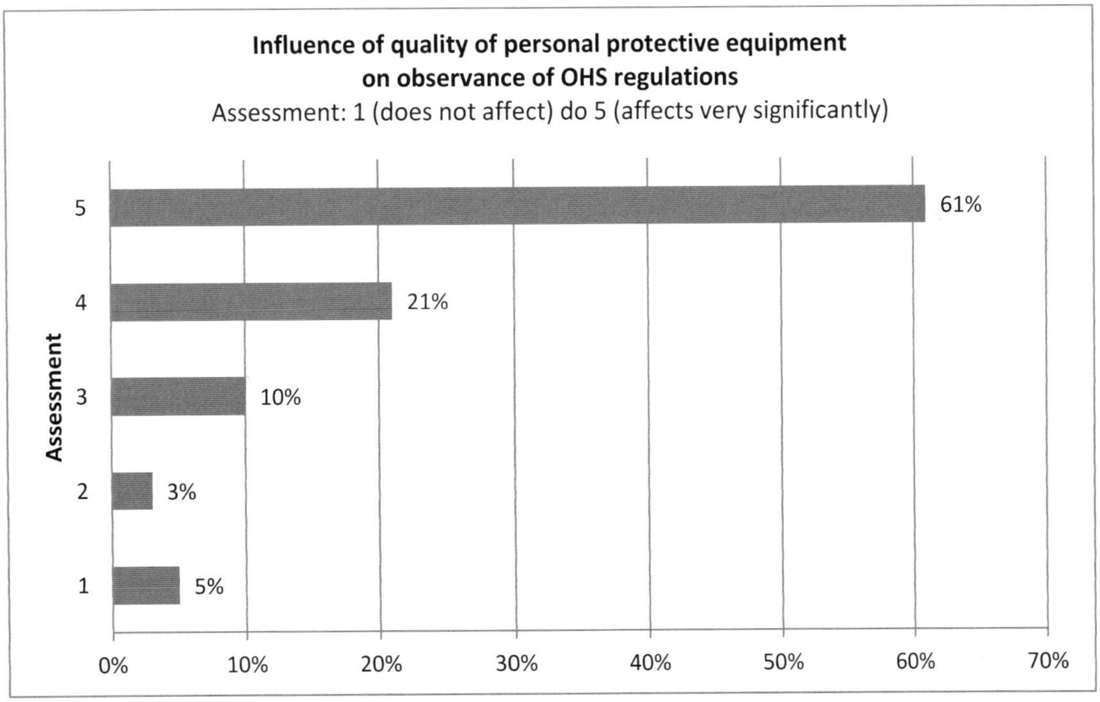

Figure 4. Influence of quality of personal protective equipment on observance of OHS regulations.

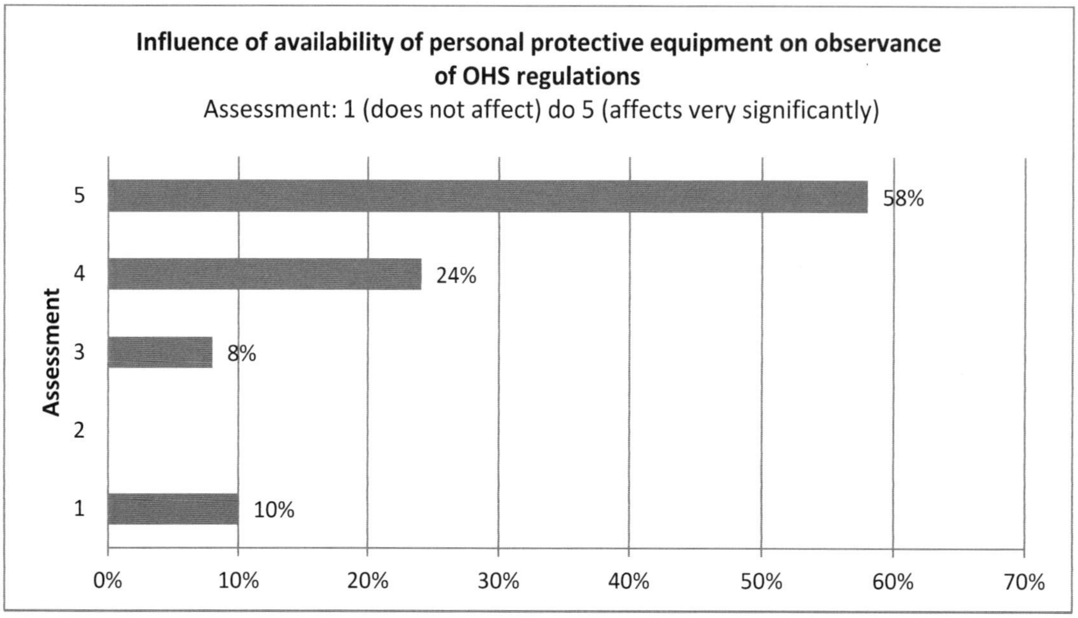

Figure 5. Influence of availability of personal protective equipment on observance of OHS regulations.

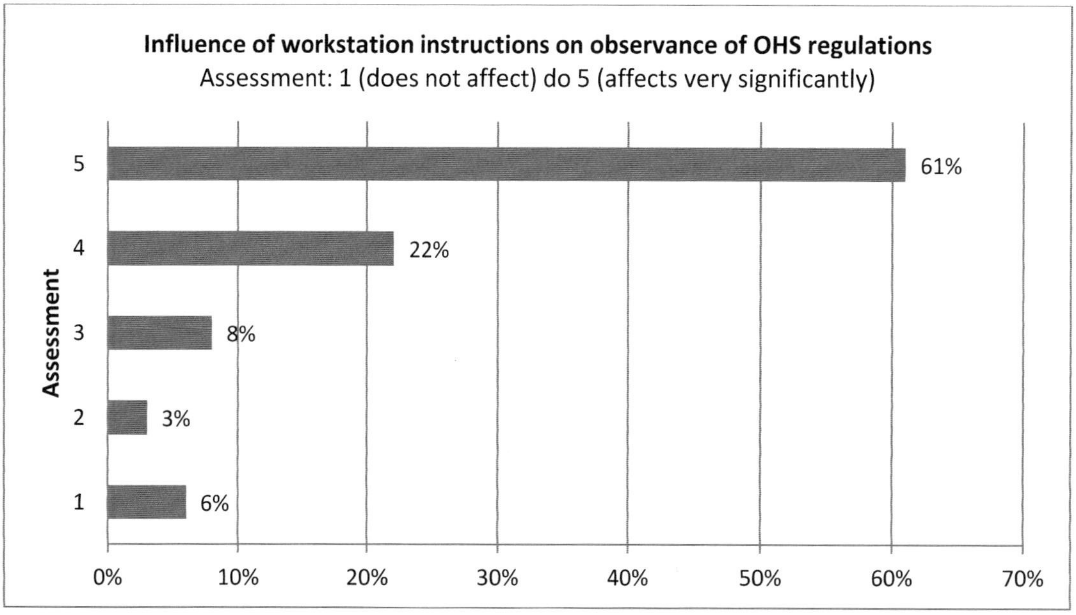

Figure 6. Influence of workstation instructions on observance of OHS regulations.

As much as 82% of the respondents thinks that personal protective equipment – their quality (figure 4) and its availability (figure 5) – to a large extent (4.0) and a very large extent (5.0) affect the observance of OHS regulations. The results of the survey indicate that a major proportion of the employees is aware the significance of personal protective equipment for work safety. The personnel of the surveyed company notice also efforts taken by the employer, who takes improving working conditions into consideration (limiting, minimising harmful effects of working environment elements) through, among other, legible workstation instructions, which according to more than 80% of the respondents affect the observance of OHS regulations to a large and a very large extent.

4. Summary of survey results

Summing up the results of the above survey it can be concluded that the presented working environment elements affect significantly the observance of OHS regulations by the employees of the chosen heavy industry production company, in which, what must be stressed, there are numerous hazards and exposures at the surveyed workstations. At the same time the survey results also indicate high awareness of the significance of the influence of the working environment quality on safe behaviour and the observance of OHS regulations. It can also show that a safety culture level, as indicated by the subject literature [12], is clearly reflected in employees' awareness and behaviour in respect of occupational health and safety.

As a result of the respondents' answers (figures 1, 2, 3, 4, 5, 6) the following conclusions can be reached:

- in the opinion of 68% of the respondents applied physical workload – energy expenditure, affect the observance of OHS regulations to a large and a very large extent (4.0 and 5.0 answers). Therefore it can be stated that a vast majority of the respondent is aware that excessive workload can cause accidents or near misses,
- as much as 82% of the respondents thinks that personal protective equipment as well as its availability affect the observance of OHS regulations to a large and a very large extent (4.0 and 5.0 answers), which can indicate that a major proportion of the surveyed company's employees understands its significance for health protection,
- however, the working environment element that affects the observance of OHS regulations workstation instructions. As much as 83% of the respondents thinks that this element affects the observance of OHS regulations to a large and a very large extent (4.0 and 5.0 answers).

The results of the presented survey show that creating appropriate working environment affects occupational health and safety not only in the area of optimisation of protection against hazards, but also building desired employee's behaviour.

References

[1] Bodak A and Gableta M 2015 Oczekiwania pracowników związane z bezpiecznym funkcjonowaniem w środowisku pracy *Zeszyty Naukowe Politechniki Śląskiej seria organizacja i zarządzanie* **83** pp 45–54

[2] Tobór-Osadnik K and Wyganowska M 2016 Budowa karty audytu behawioralnego w obszarze BHP dla wybranego stanowiska w przedsiębiorstwie górniczym *Wiadomości Górnicze* **4** pp 290–5

[3] Wilpert B and Fahlbruch B 2014 Safetyculture: Analysis and intervention ed Spitzer C, Schmocker U and Dang V N *Probabilistic safety assessment and management* Springer London **2** pp 843–9

[4] A review of safety culture and safety climate literature for the development of the safety culture inspection toolkit *Prepared by Human Engineering for the Health and Safety Executive* 2005 Bristol

[5] Bugajska J 2000 Ocena obciążenia pracą fizyczną dynamiczną na stanowisku pracy ed Koradecka D *Nauka o pracy bezpieczeństwo, higiena, ergonomia* (Warszawa: CIOP)

[6] Ejdys J, Lulewicz and A Obolewicz J 2008 *Zarządzanie bezpieczeństwem w przedsiębiorstwie (*Białystok: Wydawnictwo Politechniki Białostockiej)

[7] Matuszek J and Byrska K 2012 Ocena uciążliwości pracy na wybranych stanowiskach roboczych na podstawie określania wydatku energetycznego *Innowacje w zarządzaniu i inżynierii produkcji* (Opole: Oficyna Wydawnicza Polskiego Towarzystwa Zarządzania Produkcją) pp 733–44

[8] Wyściślik A and Gajdzik B 2010 *Jakość, środowisko i bezpieczeństwo pracy w zarządzaniu przedsiębiorstwem* (Gliwice: Wydawnictwo Politechniki Śląskiej)

[9] Prędecka A, Węgrzyn M, Sowa M and Sulak E 2016 Ocena wydatku energetycznego na stanowisku pracy piekarza metodą chronometrażowo-tabelaryczną wg Lehmanna *Zeszyty Naukowe SGSP* **59/3/** pp 137–55

[10] Pęciłło M 2011 Programy modyfikacji zachowań niebezpiecznych – wyniki wdrażania w wybranych polskich przedsiębiorstwach *Bezpieczeństwo Pracy – Nauka i Praktyka* **5** pp 9–11
[11] Ejdys J (ed) 2010 *Kształtowanie kultury bezpieczeństwa i higieny pracy w organizacji (Białystok:* Oficyna wydawnicza Politechniki Białostockiej)
[12] Rakowska A (ed) 2013 *Kultura bezpieczeństwa w przedsiębiorstwie* (Warszawa: CeDeWu.pl)